The Information Retrieval Series

Volume 47

Information Retrieval (IR) deals with access to and search in mostly unstructured information, in text, audio, and/or video, either from one large file or spread over separate and diverse sources, in static storage devices as well as on streaming data. It is part of both computer and information science, and uses techniques from e.g. mathematics, statistics, machine learning, database management, or computational linguistics. Information Retrieval is often at the core of networked applications, web-based data management, or large-scale data analysis.

The Information Retrieval Series presents monographs, edited collections, and advanced text books on topics of interest for researchers in academia and industry alike. Its focus is on the timely publication of state-of-the-art results at the forefront of research and on theoretical foundations necessary to develop a deeper understanding of methods and approaches.

This series is abstracted/indexed in EI Compendex and Scopus.

Andrea Esuli • Alessandro Fabris •
Alejandro Moreo • Fabrizio Sebastiani

Learning to Quantify

 Springer

Andrea Esuli
Istituto di Scienza e Tecnologie
dell'Informazione
Consiglio Nazionale delle Ricerche
Pisa, Italy

Alessandro Fabris
Dipartimento di Ingegneria
dell'Informazione
Università di Padova
Padova, Italy

Alejandro Moreo
Istituto di Scienza e Tecnologie
dell'Informazione
Consiglio Nazionale delle Ricerche
Pisa, Italy

Fabrizio Sebastiani
Istituto di Scienza e Tecnologie
dell'Informazione
Consiglio Nazionale delle Ricerche
Pisa, Italy

This work was supported by Istituto di Scienza e Tecnologie dell'Informazione

ISSN 1871-7500   ISSN 2730-6836   (electronic)
The Information Retrieval Series
ISBN 978-3-031-20466-1   ISBN 978-3-031-20467-8   (eBook)
https://doi.org/10.1007/978-3-031-20467-8

This Springer imprint is published by the registered company Springer Nature Switzerland AG
The registered company address is: Gewerbestrasse 11, 6330 Cham, Switzerland

Policy makers or computer scientists may be interested in finding the needle in the haystack (...), but social scientists are more commonly interested in characterizing the haystack.
(Daniel J. Hopkins and Gary King, 2010)

Preface

In a number of applications involving classification, the final goal is not determining which class (or classes) individual unlabelled instances belong to, but estimating the *prevalence* (or "relative frequency", or "prior probability") of each class in the unlabelled data. In recent years it has been pointed out that, in these cases, it would make sense to directly optimise machine learning algorithms for this goal, rather than (somehow indirectly) just optimising the classifier's ability to label individual instances. The task of training estimators of class prevalence via supervised learning is known as *learning to quantify*, or, more simply, *quantification*. It is by now well known that performing quantification by classifying each unlabelled instance via a standard classifier and then counting the instances that have been assigned to the class (the *Classify and Count* method) usually leads to biased estimators of class prevalence, i.e., to poor quantification accuracy; as a result, methods (and evaluation measures) that address quantification as a task in its own right have been developed. This book covers the main applications of quantification, the main methods that have been developed for learning to quantify, the measures that have been adopted for evaluating it, and the challenges that still need to be addressed by future research.

The book is divided in seven chapters. Chapter 1 sets the stage for the rest of the book by introducing fundamental notions such as class distributions, their estimation, and dataset shift, by arguing for the suboptimality of using classification techniques for performing this estimation, and by discussing why learning to quantify has evolved as a task of its own, rather than remaining a by-product of classification. Chapter 2 provides the motivation for what is to come by describing the applications that quantification has been put at, ranging from improving classification accuracy in domain adaptation, to measuring and improving the fairness of classification systems with respect to a sensitive attribute, to supporting research and development in the social sciences, in political science, epidemiology, market research, and others. In Chapter 3 we move on to discuss the experimental evaluation of quantification systems; we look at evaluation measures for the various types of quantification systems (binary, single-label multiclass, multi-label multiclass, ordinal), but also at evaluation protocols for quantification, that essentially consist in ways to extract multiple testing samples for use in quantification evaluation from a single classification test set. Chapter 4 is possibly the central chapter of the book, and looks at the various supervised learning methods for learning to quantify that have been proposed over the years, be they of an aggregative nature (i.e., methods that require the classification of all individual unlabelled items as an intermediate step) or of a non-aggregative nature (i.e., methods in which no classification of individual items is performed). In Chapter 5 we look at a number of "advanced" (or niche) topics in quantification, including quantification for ordinal data, cross-lingual quantification of textual items, quantification for networked data, and quantification for streaming data. Chapter 6 looks at other aspects of the "quantification landscape" that have not been covered in the previous chapters, and discusses the evolution of quantification research, from its beginnings to the most recent quantification-based "shared tasks", the landscape of quantification-based, publicly available software libraries, and other tasks in data science that present important similarities with quantification. Chapter 6 also presents the results of experiments, that we have carried out ourselves, in which we compare many of the methods discussed in Chapter 4 on a common testing infrastructure. Chapter 7 concludes the book, pointing to potential future developments in the quantification arena.

The book is mostly addressed to researchers in data science that might want to come up to speed with the state of the art in learning to quantify, but it can be useful also to researchers and scientists that operate in other disciplines and that apply techniques from data science to their own application domains. Indeed, it is our experience that many potential users of quantification techniques (who operate in the fields touched upon in Chapter 2, and possibly in others too) do not use them, thus settling for suboptimal "classify and count" techniques, for the simple fact that they are not aware of their existence, and of the existence of quantification as a task of its own; it is also those potential users that we hope will be inspired by this book.

We thus hope that the availability of a book that surveys all aspects of the quantification workflow and presents them in a hopefully accessible form, will increase the interest in this subject on the part of researchers and practitioners alike, and will contribute to making quantification better known to potential users of this technology and to researchers interested in advancing the field.

Pisa, Italy   Andrea Esuli
Padova, Italy   Alessandro Fabris
Pisa, Italy   Alejandro Moreo
Pisa, Italy   Fabrizio Sebastiani

Acknowledgments

The work of Andrea Esuli, Alejandro Moreo, and Fabrizio Sebastiani has been supported by the SoBigData++ project, funded by the European Commission (Grant 871042) under the H2020 Programme INFRAIA-2019-1, by the AI4Media project, funded by the European Commission (Grant 951911) under the H2020 Programme ICT-48-2020, and by the SoBigData.it and FAIR projects, funded by the Italian Ministry of University and Research under the NextGenerationEU program. The authors' opinions do not necessarily reflect those of the European Commission. The work by Alessandro Fabris was supported by MIUR (Italian Ministry for University and Research) under the "Departments of Excellence" initiative (Law 232/2016).

Contents

Acronyms

Chapter 1
The Case for Quantification

Classification, perhaps the most fundamental among the tasks addressed by supervised machine learning, has to do with assigning one or more classes from a predefined set to each data item from a given distribution. Over the last 50 years or more, classification has been extensively studied, not only in machine learning but also in philosophy, content analysis, statistics, and other branches of science.

About fifteen years ago, in a seminal paper, Forman (2005) observed that, in several applications involving classification, the final goal is not determining which class (or classes) individual unlabelled data items belong to, but estimating the *prevalence* (also called "relative frequency", or "prior probability", or simply "prior") of each class in the unlabelled data. Training class prevalence estimators via supervised learning has come to be known as *quantification*, a term coined by Forman (2005, 2006, 2008) which has stuck from then on; the term *learning to quantify* is also used, which stresses the fact that prevalence estimation is, in this case, tackled by means of supervised learning.

To see the importance of learning to quantify, let us examine the task of classifying textual answers returned to open-ended questions in questionnaires (Esuli and Sebastiani, 2010b), and let us discuss two important such scenarios.

In the first scenario, a telecommunications company asks its current customers the question "How satisfied are you with our mobile phone services?", and has its information scientists classify each resulting textual answer into one of a set of classes of interest. One of the goals of this survey is to know which of the resulting textual answers belong to class MayDefectToCompetition. The company is likely interested in accurately classifying each individual customer, since it may want to call each customer that is assigned the class MayDefectToCompetition and offer her improved conditions, so as not to lose her as a customer.

In the second scenario, a market research expert, working for a fast food company, asks respondents the question "What do you think of onions in cheeseburgers?", and wants to know which of the resulting textual answers belong to class LikesOnionsInCheeseburgers. Here, the market research expert is presumably

© The Author(s) 2023
A. Esuli et al., *Learning to Quantify*, The Information Retrieval Series 47,
https://doi.org/10.1007/978-3-031-20467-8_1

not interested in whether a specific individual belongs to the class, but is likely interested in knowing *how many* respondents, out of the total number of respondents, belong to it, i.e., in knowing the prevalence of the class.

In sum, while in the former scenario the interest is at the individual level, in the latter the aggregate level is all that matters; in other words, in the former scenario classification is the goal, while in the latter the real goal is quantification.

Other tasks in which "individuals do not matter", i.e., in which the classes to which belong are useful only inasmuch as they allow us to obtain indicators concerning the entire population, are, e.g., predicting election results by estimating the prevalence of blog posts (or tweets) supporting a given candidate or party (Hopkins and King, 2010), or planning the amount of human resources to allocate to different types of issues in a customer support centre by estimating the prevalence of customer calls related to each issue (Forman, 2005), or supporting epidemiological research by estimating the prevalence of medical reports where a specific pathology is diagnosed (Baccianella et al., 2013). Indeed, there are entire fields of human inquiry which are devoted to studying phenomena only at a collective level; examples of such fields are market research, political science, the social sciences, ecological modelling, and epidemiology. When researchers in these fields are confronted with unlabelled data and the need to label them, they usually need quantification, and not classification.

Note that, also due to the variety of fields in which it has emerged as an application need, quantification goes under different names, in different areas of science and in different scientific papers. It has variously been called *counting*, (Lewis, 1995), *class probability re-estimation* (Alaíz-Rodríguez et al., 2011), *class prior estimation* (Chan and Ng, 2006; Zhang and Zhou, 2010), and *class distribution estimation* (González-Castro et al., 2013; Limsetto and Waiyamai, 2011; Xue and Weiss, 2009).

1.1 Class Distributions and Their Estimation

An example quantification task is displayed, via a histogram, in Figure 1.1. The example involves a number of textual product reviews labelled according to a set of five classes (from VeryNegative to VeryPositive) representing "scores" assigned to the reviewed products. In the histogram, the blue bars represent the true (unknown) class prevalence values that need to be estimated (i.e., the fractions of product reviews that have been assigned the scores indicated), and the red bars represent the corresponding estimates obtained by a quantification method. When the blue bars are identical to the corresponding red bars, the estimation is perfectly accurate. Since all the fractions are in the [0,1] interval and sum up to 1, we are here in the presence of two *probability distributions*. This shows that learning to quantify may be also defined as the task of learning to approximate an unknown *true distribution* by a *predicted distribution*. (In the case of Figure 1.1 we are actually in the presence of two *ordinal* distributions, since there are more than two classes and there is

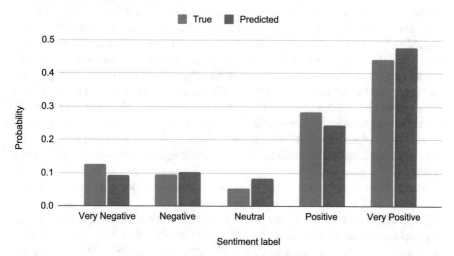

Fig. 1.1 An example quantification task; blue bars represent the unknown true class prevalence values that need to be estimated, and red bars represent their estimates obtained by a quantification method.

an implied total order on them; see Sections 3.2 and 5.1 for more on ordinal distributions and their estimation.) As a result, and as we will see more thoroughly in Section 3, practically all evaluation measures for quantification are *divergences*, i.e., measures of how a predicted distribution "diverges" from the true distribution. This justifies the fact that, as previously hinted, quantification is sometimes called "class distribution estimation" (González-Castro et al., 2013; Limsetto and Waiyamai, 2011; Xue and Weiss, 2009).

1.2   The Suboptimality of *Classify and Count*

In the absence of methods for estimating class prevalence values more directly, the obvious method for doing it is *Classify and Count*, i.e., classifying each unlabelled data item and estimating class prevalence values by counting the items that have been assigned to each class.

However, this strategy is sub-optimal: while a perfect classifier is also, quite obviously, a perfect *quantifier* (i.e., estimator of class prevalence values), a good classifier may be a bad quantifier. To see this, one only needs to look at the definition of F_1, a standard evaluation function for binary classification, which is defined as

$$F_1 = \frac{2TP}{2TP + FP + FN} \tag{1.1}$$

where TP, FP, FN indicate the numbers of true positives, false positives, and false negatives, respectively, in a binary contingency table. According to F_1, a binary classifier h_1 for which FP $=$ 10 and FN $=$ 10 is worse than a classifier h_2 for which, on the same test set, FP $=$ 8 and FN $=$ 10. However, when using "classify and count", h_1 is intuitively a better binary quantifier than h_2; indeed, h_1 is (on this test set) a perfect estimator of class prevalence values, since FP and FN are equal and thus compensate each other, so that the distribution of the unlabelled items across the class and its complement is estimated perfectly. That a good classifier may be a bad quantifier can be seen by the fact that, as evident from Equation 1.1, F_1 considers "good" those classifiers that keep the sum (FP $+$ FN) to a minimum; however, the goal of a quantification algorithm must be that of keeping to a minimum $|$FP $-$ FN$|$, and not (FP $+$ FN).

The above example shows that even an accurate classifier may be *biased*, i.e., may keep its false positives to a minimum only at the expense of a substantially higher number of false negatives (or vice versa); if this is the case, the classifier is a bad quantifier. This phenomenon is not infrequent, especially in the presence of imbalanced data, i.e., data in which the items from the majority class by far outnumber the items from the other classes. This is very frequent, say, in text classification, where data relevant to a certain topic are often a tiny fraction of the entire set; but occurs in all other contexts in which the amount of "signal" is much smaller than the amount of "noise". In such cases, learning algorithms that minimise "standard" loss functions (i.e., the Hamming loss, the hinge loss, or their proxies) often generate classifiers with a tendency to choose the majority class, which means a much higher number of false positives than false negatives for the majority class, which means in turn that such an algorithm will tend to underestimate the counts of minority classes. For instance, Esuli and Sebastiani (2015) report an experimentation on 5,148 binary test sets averaging 15,000+ examples each, in which a linear SVM delivers an average FN$/$FP ratio of 0.109 for the majority class; by contrast, for a perfect estimator of class prevalence values this ratio is 1.

The previous arguments indicate that *quantification should not be considered a mere by-product of classification, and should be studied and solved as a task of its own*. There are at least two other arguments that support this idea. One is that the functions that are used for evaluating classification cannot be used for evaluating quantification, since these functions measure, by and large, how many data items have been misclassified, and not how much the estimated class prevalence values differ from the true class prevalence values. This means that the learning algorithms that minimise these functions are optimised for classification, and not for quantification. (We will come back on this topic in Section 4.3.1.) A second, symmetrical argument, put forth by Forman (2008), is that methods specifically devised for learning to quantify require fewer training data in order to deliver the same quantification accuracy as standard methods based on "classify and count". While Forman's observation is of an empirical nature, there are also theoretical arguments that support this fact, which will be more thoroughly discussed in Section 4.4.

1.3 Notational Conventions

Since in the next section we will start discussing quantification in some mathematical detail, we now fix some notation. By $\mathbf{x}$ we will indicate a data item drawn from a domain $\mathcal{X}$, represented as a vector of features. By y we will indicate a class drawn from a set of classes (or *codeframe*) $\mathcal{Y} = \{y_1, \ldots, y_{|\mathcal{Y}|}\}$, and by $\overline{y}$ we will indicate its complement, i.e., $\overline{y} = \bigcup_{y_i \in \mathcal{Y} \setminus \{y\}} y_i$. When the codeframe contains just two classes we will often indicate this codeframe as $\mathcal{Y} = \{\oplus, \ominus\}$, and will call $\oplus$ "the positive class" and $\ominus$ "the negative class". Given $\mathbf{x} \in \mathcal{X}$ and $y \in \mathcal{Y}$, a pair $(\mathbf{x}, y)$ will thus denote a data item with its class label; given a pair $(\mathbf{x}, y)$ we will also write $\Phi(\mathbf{x}) = y$, i.e., $\Phi(\mathbf{x})$ will indicate the label of $\mathbf{x}$.[1] The symbol σ will denote a *sample*, i.e., a non-empty set of (labelled or unlabelled) items drawn from $\mathcal{X}$. Given a class y_i, we will denote by σ_i the set of items in sample σ that belong to y_i; we will denote by $|\sigma|$ the number of items contained in σ.

By $p_\sigma(y)$ we will indicate the true prevalence of class y in sample σ, by $\hat{p}_\sigma(y)$ we will indicate an estimate of this prevalence[2], and by $\hat{p}_\sigma^M(y)$ we will indicate the estimate of this prevalence as obtained via quantification method M. In other words, symbol p will denote a true distribution of the unlabelled items across codeframe $\mathcal{Y}$, while symbol $\hat{p}$ will denote a predicted distribution (or *estimator*), i.e., the result of estimating an unknown true distribution; symbol $\mathcal{P}$ will denote the (infinite) set of all distributions on $\mathcal{Y}$.[3] By $D(p, \hat{p})$ we will denote an evaluation measure for quantification.

A sample of labelled items (that we will typically use as a training set) will be denoted by L, while a sample of unlabelled items (that we will typically use as a sample to quantify on) will be denoted by U.

We will take a *(hard) classifier* to be a function $h : \mathcal{X} \to \mathcal{Y}$. By $p_\sigma^h(\hat{y})$ we will denote the prevalence in sample σ of the data items that have been assigned to class y by classifier h. When dealing with binary contexts, we will use TP, FP, FN, TN, to denote the numbers of true positives, false positives, false negatives, true negatives, respectively, as resulting from the application of a hard classifier to an unlabelled sample U, and as contained in the resulting binary contingency table.

We will instead take a *soft classifier* to be a function $s : \mathcal{X} \to [0, 1]^{|\mathcal{Y}|}$ such that each $s(\mathbf{x})$ is a vector of $|\mathcal{Y}|$ *posterior probabilities* (each indicated as $p(y|\mathbf{x})$) and such that $\sum_{y \in \mathcal{Y}} p(y|\mathbf{x}) = 1$; $p(y|\mathbf{x})$ indicates the probability of membership in y

---

[1] For the moment being we assume that a data item $\mathbf{x} \in \mathcal{X}$ can belong to one and only one class $y \in \mathcal{Y}$; the reason for this will be explained in Section 1.4.

[2] Consistently with most mathematical literature, we use the caret symbol (ˆ) to indicate estimation.

[3] In order to keep things simple we avoid overspecifying the notation, thus leaving some aspects of it implicit; e.g., in order to indicate a true distribution p of the unlabelled items in a sample σ across a codeframe $\mathcal{Y}$ we will often write p instead of the more cumbersome $p_\sigma^{\mathcal{Y}}$, thus letting σ and $\mathcal{Y}$ be inferred from context.

Table 1.1 Notation for the symbols most frequently used in this book.

Symbol	Meaning
$\mathbf{x} \in \mathcal{X}$	A data item from domain $\mathcal{X}$
$y, \overline{y} \in \mathcal{Y}$	A class from codeframe $\mathcal{Y}$ and its complement in $\mathcal{Y}$
$\oplus, \ominus \in \mathcal{Y}$	The two classes of a binary codeframe $\mathcal{Y}$
$\Phi(\mathbf{x}) \in \mathcal{Y}$	The class label of data item $\mathbf{x}$
σ	A sample of data items drawn from domain $\mathcal{X}$
$|\sigma|$	Cardinality of sample σ
L	A labelled sample of items (typically: for training purposes)
U	An unlabelled sample of items (typically: for testing purposes)
$p_\sigma(y)$	True prevalence of class y in σ
$\hat{p}_\sigma(y)$	Estimate of the prevalence of class y in σ
$\hat{p}_\sigma^M(y)$	Estimate $\hat{p}_\sigma(y)$ obtained via method M
$p_\sigma^h(\hat{y})$	Fraction of elements in sample σ to which h has assigned class y
$D(p, \hat{p})$	An evaluation measure for the prevalence estimate
$h : \mathcal{X} \rightarrow \mathcal{Y}$	A hard classifier for $\mathcal{Y}$
$s : \mathcal{X} \rightarrow [0, 1]^{|\mathcal{Y}|}$	A soft classifier for $\mathcal{Y}$
$p(y|\mathbf{x})$	"Posterior" probability that $\mathbf{x}$ is in y

of item $\mathbf{x}$ as estimated by s.[4] A hard classifier is obtained from a soft classifier by taking

$$h(\mathbf{x}) = \arg\max_{y \in \mathcal{Y}} p(y|\mathbf{x}) \tag{1.2}$$

Table 1.1 summarises these symbols for convenience.

1.4   Quantification Problems

Similarly to classification, learning to quantify admits different problems of applicative interest, based (a) on how many classes codeframe $\mathcal{Y}$ contains, and (b) how many of the classes in $\mathcal{Y}$ can be attributed at the same time to the same item. We characterise quantification problems as follows:

1. *Single-Label Quantification* (SLQ) is defined as quantification when each data item belongs to exactly one of the classes in $\mathcal{Y} = \{y_1, \ldots, y_{|\mathcal{Y}|}\}$.

---

[4] Another way of saying this is that s is a function that maps the domain $\mathcal{X}$ onto the *probability simplex* (aka *standard simplex*) $\Delta^{|\mathcal{Y}|}$, defined as the unit $(|\mathcal{Y}| - 1)$-simplex.

2. *Multi-Label Quantification* (MLQ) is defined as quantification when the same item may belong to any number of classes (zero, one, or several) in $\mathcal{Y} = \{y_1, \ldots, y_{|\mathcal{Y}|}\}$.
3. *Binary Quantification* (BQ) may alternatively be defined

   (a) as SLQ with $|\mathcal{Y}| = 2$ (in this case $\mathcal{Y} = \{y_1, y_2\}$ (or, as we will often write in the binary case, $\mathcal{Y} = \{\oplus, \ominus\}$) and each item must belong to either y_1 or y_2), or

   (b) as MLQ with $|\mathcal{Y}| = 1$ (in this case $\mathcal{Y} = \{y\}$ and each item either belongs or does not belong to y).

4. *Ordinal Quantification* (OQ) is defined as single-label quantification when the codeframe $\mathcal{Y} = \{y_1, \ldots, y_{|\mathcal{Y}|}\}$ is such that there exists a total order $y_1 \prec \ldots \prec y_{|\mathcal{Y}|}$ among its classes. (The example discussed in Section 1.1 was of this type.)
5. *Regression Quantification* (RQ), a task which stands to regression as "standard" quantification stands to classification. This task slightly falls outside the characterisation of "quantification" that we have given in the previous sections, since there is no set $\mathcal{Y}$ of classes involved, i.e., each item is labelled with a real-valued score and quantification amounts to estimating the fraction of items whose score is in a given interval $[a, b]$, with $a, b \in \mathbb{R}$.

Among the above tasks, the one we will mostly devote our attention to in this book is SLQ. The reasons for doing this are the following:

- BQ is a special case of SLQ (see Bullet 3a above), which means that any method for performing SLQ and any measure for evaluating SLQ can also be used for BQ.
- Any quantification method for BQ can also be used for MLQ, since MLQ can be solved by deploying $|\mathcal{Y}|$ independent binary quantification systems, one for each $y \in \mathcal{Y}$.[5] Additionally, any evaluation measure for BQ can be used for evaluating MLQ, since MLQ can be evaluated by checking, for each $y \in \mathcal{Y}$, how well $\hat{p}(y)$ approximates $p(y)$ by means of an evaluation measure for BQ that uses $\{y, \overline{y}\}$ as the binary codeframe.
- Most quantification methods and evaluation measures that have been proposed in the literature were either proposed for SLQ, or were originally proposed for BQ and can be easily extended to SLQ. Conversely, there has been very little work on methods for solving OQ or RQ, and on measures for evaluating them.

---

[5] MLQ might in principle be solved in ways other than by recasting the problem into $|\mathcal{Y}|$ independent binary quantification problems, i.e., it might be solved by attempting to leverage possible stochastic dependencies between the classes in $\mathcal{Y}$, similarly to what is done in many approaches to multi-label classification. For MLQ, the only attempt we are aware of from past literature is by Levin and Roitman (2017). However, in this work the problem is tackled as a set of independent binary quantification problems, and the correlations among the classes are never brought to bear.

However, while SLQ will be the main focus of the book, the solutions that have been proposed in the literature for other quantification problems, such as OQ and RQ, will also be discussed.

1.5   Dataset Shift and Quantification

Standard supervised learning algorithms are based on the assumption that the training data and the unlabelled data the predictor is supposed to issue predictions about (which in experimental settings is represented by the test set), are *independently and identically distributed* (IID). In other words, since labelled data items are represented by pairs of type $(\mathbf{x}, y)$, the distribution of pairs in the labelled set is assumed to be the same as that on the set of unlabelled items, i.e., $p_L(\mathbf{x}, y) = p_U(\mathbf{x}, y)$. Of particular interest to quantification is the fact that, *a fortiori*, the distribution of labels is assumed to stay constant, i.e., $p_L(y) = p_U(y)$.

But the world we live in and the data it provides are constantly evolving, and the scenarios in which we might want to deploy the trained models may widely differ. For instance, in an effort to use quantification technology for estimating the prevalence of different species of living beings on the seabed (see Figure 1.2),

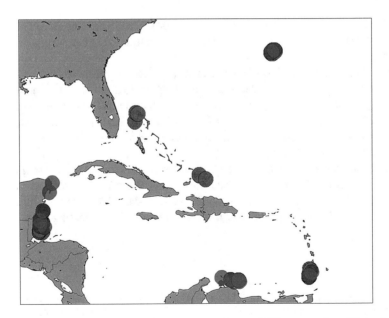

Fig. 1.2 Using quantification for estimating the prevalence of different species of living beings on the seabed; red circles indicate the locations where the training data were collected while blue circles indicate the locations where the unlabelled data to which the trained model was applied were collected (from Beijbom et al., 2015).

Beijbom et al. (2015) train a model on labelled data mostly collected on the coasts off the Bahamas and Caicos islands, and apply the trained model on unlabelled data collected in various other locations, including the coasts off Mexico and Venezuela. In this case, and in many other cases, the IID assumption is violated, and $p_L(\mathbf{x}, y) \neq p_U(\mathbf{x}, y)$; this phenomenon is usually referred to as *dataset shift* (Moreno-Torres et al., 2012; Quiñonero-Candela et al., 2009).[6] Of interest to us is the particular case in which $p_L(y) \neq p_U(y)$, which is often called *distribution shift* (Bella et al., 2014).

Example reasons why distribution shift may occur are the following:

1. The environment might not be stationary across time and/or space and/or other variables, which means that the testing conditions would be irreproducible at training time. For instance, in a backlog of newswire stories from year 2001, the prevalence of class Terrorism in news produced before September 11 will not be the same as in news produced after September 11; likewise, the prevalence of reviews scored as VeryPositive for a certain product will not be the same before and after its price has been cut in half. These are cases of non-stationarity across time. Cases of non-stationarity across space are also ubiquitous: for instance, the same news stories that fall under class HomeNews in the UK fall instead under class Europe in the US, and the relative frequencies of classes Cricket and Baseball are likely very different in datasets of news originating from these two countries. The case illustrated in Figure 1.2 is also a case of non-stationarity across space.

2. The process of labelling training data might be class-dependent. For instance, assume we need a predictor that recognises the presence of a rare disease (e.g., a 1-in-a-million-cases disease). A training set consisting of one positive example and 999,999 negative examples would likely deliver a highly ineffective classifier, so we will typically try to insert into the training set as many positive examples as we can put our hands on, and to remove from it a sizeable number of negative examples. This means that the prevalence of the minority class in the training set L will be much higher than its prevalence in the set U of unlabelled items, and that (symmetrically) the prevalence of the majority class in L will be much lower than its prevalence in U.

3. The labelling process might introduce bias in the training set. For instance, suppose we build our training set via active learning. The active learning process might ask the human assessor to annotate items that have a high probability of membership in the minority class (this is indeed the stated goal of the "relevance sampling" active learning technique Lewis and Gale, 1994), since training examples of the minority class are usually more informative than those of the majority class. Also in this case the prevalence of the minority class in the training set will be much higher than its prevalence in the unlabelled set. More in

---

[6] The word "drift" is also often used in place of "shift" in the machine learning literature; this applies not only to term "dataset shift " but also to the various types of shift we will discuss in this section.

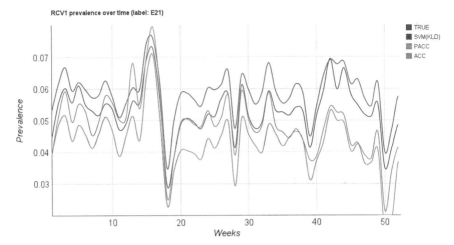

Fig. 1.3   Example of distribution shift in the RCV1-v2 test collection.

general, when using active learning in order to build the training set, dataset shift
will be present regardless of the active learning technique used (i.e., relevance
sampling or other), for the simple fact that all active learning techniques force
the assessor to annotate items in a non-random fashion, and this divergence from
randomness inherently means dataset shift.

Bullets 2 and 3 are both examples of *sample selection bias*, a term that refers to the
presence of a systematic bias (sometimes intended, sometimes unintended) either in
the process of data collection or in the process of data labelling, and to the fact that
due to this bias the distribution of training examples ends up being different from
the distribution of data in the domain to be modelled.

Figure 1.3 illustrates an example of distribution shift in the well-known RCV1-
v2 test collection (Lewis et al., 2004). This collection consists of one year's worth
of timestamped news published by Reuters from Aug 20, 1996, to Aug 19, 1997.
The blue curve in Figure 1.3 is the result of binning these 804,414 news stories
into 52 bins, one per week, and computing the prevalence in each bin of one of the
101 classes (class E21) of which the codeframe consists. The x axis indicates the
week on which the prevalence value is to be computed, while the y axis indicates
the corresponding prevalence value. The blue curve represents the true prevalence
values, while the other three curves represent the prevalence values as estimated by
three of the quantification methods that we will discuss in Section 4 (the ACC,
PACC, and SVM(KLD) methods, discussed in Sections 4.2.3, 4.2.4, and 4.3.1,
respectively). The fact that there is distribution shift in this dataset is shown by
the fact that the blue curve is not a flat line; a good quantifier is one that generates
a curve as close to the blue curve as possible. Note that, while there is indeed some
dataset shift here, its magnitude is not high, as shown by the fact that the oscillations

of the blue curve around the $y = 0.05$ line are moderate. Other applicative scenarios exhibit a much more marked distribution shift.

Note that the presence of dataset shift, and of distribution shift in particular, is the *raison d'être* of applications that track class prevalence across different contexts (i.e., across time, space, or other variables), i.e., of studying quantification. If we could assume that there is no dataset shift, i.e., that $p_L(\mathbf{x}, y)$ always equals $p_U(\mathbf{x}, y)$, the optimal quantification strategy would be to assume that, for each $y \in \mathcal{Y}$, $p_L(y) = p_U(y)$ for all unlabelled samples U. (This trivial strategy, that we call *Maximum Likelihood Prevalence Estimation* (MLPE), will be discussed in Section 4.1.) In other words, the reason for studying and solving quantification lies in the awareness that dataset shift, and distribution shift in particular, exists.

1.5.1 Types of Dataset Shift and Their Relation to Quantification

In order to assess the impact of distribution shift on quantification, it is useful to note that $p(y)$ may be written as

$$p(y) = \sum_{\mathbf{x}} p(y|\mathbf{x}) p(\mathbf{x}) \qquad (1.3)$$

When any of $p(y|\mathbf{x})$ and $p(\mathbf{x})$ vary in switching from the training data to the unlabelled data, distribution shift occurs. The case in which $p(\mathbf{x})$ varies occurs when certain regions of the feature space are more densely populated in U than in L while other regions are correspondingly less densely populated in U than in L; this phenomenon is usually called *covariate shift*. For instance, the example about class Terrorism in Bullet 1 above is a case of covariate shift, as is the example in Bullet 2. Instead, the case in which $p(y|\mathbf{x})$ varies occurs when the meaning of class y has changed (where "meaning" is to be understood in the sense of extensional semantics), and the very same item $\mathbf{x}$ that had label y in L might not have label y in U; this phenomenon is usually called *concept shift*. For instance, the example about news falling in the HomeNews or Europe classes in Bullet 1 above is a case of concept shift.

Figure 1.4 (taken from Bella et al., 2014) exemplifies covariate shift, concept shift, and the distribution shift that derives from them, in graphical form. The plots are the result of an experiment for a regression task, where labels take values not on a discrete codeframe but on the set of real numbers (here: on the [0,1] interval), and where we assume the existence of a single feature x. The top left sub-figure shows the distribution of the examples in the training set. The top right sub-figure shows the distribution of the examples in a test set which exhibits neither covariate shift nor concept shift (i.e., the training set and the test set are IID). The bottom left sub-figure shows the distribution of the examples in a test set which exhibits no concept shift (since $p(y|\mathbf{x})$ is the same as in the training set) but reveals the presence

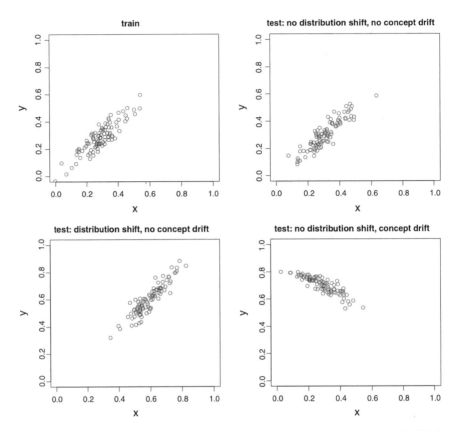

Fig. 1.4 Distribution shift and concept shift in regression; the image is from Bella et al. (2014), where "concept shift" is called (as often happens in machine learning literature) "concept drift".

of covariate shift (since $p(\mathbf{x})$ is not the same as in the training set), which in turns generates distribution shift ($p(y)$ not being the same as in the training set). The bottom right sub-figure shows the distribution of the examples in a test set which exhibits no covariate shift (since $p(\mathbf{x})$ is the same as in the training set) but reveals the presence of concept shift (since $p(y|\mathbf{x})$ is not the same as in the training set), which also causes distribution shift to happen ($p(y)$ being different in the training set and in the test set).

Covariate shift, concept shift, distribution shift, and Equation 1.3 are relevant in what Fawcett and Flach (2005) have called $\mathcal{X} \rightarrow \mathcal{Y}$ *problems*, i.e., problems in which it is the values of the features in $\mathbf{x}$ that probabilistically determine the label y of $\mathbf{x}$. An example of an $\mathcal{X} \rightarrow \mathcal{Y}$ learning problem is weather forecasting, since it is a number of climatic conditions (e.g., pressure, temperature, humidity, etc., that can be represented in a feature vector $\mathbf{x}$) that determine whether it is going to snow or not (a fact that can be represented by a binary dependent variable y), and not the other way around. In these cases, if the distribution of climatic conditions shifts, the probability that it is going to snow shifts too.

It is also useful to note that $p(\mathbf{x})$ may be written as

$$p(\mathbf{x}) = \sum_{y} p(\mathbf{x}|y)p(y) \tag{1.4}$$

This equation is instead relevant in what Fawcett and Flach (2005) have called $\mathcal{Y} \rightarrow \mathcal{X}$ *problems*, i.e., problems in which the class to which a data item $\mathbf{x}$ belongs probabilistically determines the values of the features in vector $\mathbf{x}$. An example of a $\mathcal{Y} \rightarrow \mathcal{X}$ learning problem is authorship attribution, i.e., the task of inferring the author (from a set of $|\mathcal{Y}|$ candidate authors) of a text of unknown or disputed paternity (Koppel et al., 2009). Authorship attribution, a task which is usually carried out by using as features a number of "stylistic" traits that tend to characterise an author's writing style, is an $\mathcal{Y} \rightarrow \mathcal{X}$ problem, since it is the fact that a certain text is, say, Shakespeare's, that causes it to have certain stylistic characteristics, and not the other way around. In $\mathcal{Y} \rightarrow \mathcal{X}$ problems, when $p(y)$ varies across L and U, it does so "autonomously" (since y is a cause, and not an effect); this phenomenon is usually called *prior probability shift* (Storkey, 2009), or sometimes *label shift* (Alexandari et al., 2020).

In the context of text classification, Card and Smith (2018) call the class labels attached to data items in $\mathcal{X} \rightarrow \mathcal{Y}$ problems *extrinsic labels*, while they call the ones in $\mathcal{Y} \rightarrow \mathcal{X}$ problems *intrinsic labels*. The rationale of these names is that in $\mathcal{Y} \rightarrow \mathcal{X}$ problems the labels are intrinsic properties of the data item, and precede the generation of the data item itself, while this is not the case in $\mathcal{X} \rightarrow \mathcal{Y}$ problems. In other words, in $\mathcal{X} \rightarrow \mathcal{Y}$ problems, whether the label of a data item $\mathbf{x}$ is y or not is open to subjective interpretation, while it is not in $\mathcal{Y} \rightarrow \mathcal{X}$ problems.

However, it should be noted that it is not always easy to characterise with certainty a given problem as being of type $\mathcal{X} \rightarrow \mathcal{Y}$ or of type $\mathcal{Y} \rightarrow \mathcal{X}$; sometimes this question looks a bit akin to wondering which of chicken and egg came first. As a result, different types of shift (covariate shift, concept shift, prior probability shift) that concur in causing distribution shift may be at play at the same time.

In realistic settings, distribution shift is bound to happen at some scale. Its magnitude might just be negligible, in which case the performance of a classifier at deployment will be nearly unaffected, and $p_U(y)$ will be well approximated by $p_L(y)$. However, in the absence of guarantees stemming from domain expertise, a cautious approach will include a procedure to monitor distribution shift and, ideally, a mechanism to adapt to it.[7]

---

[7] An example test for checking if the class prevalence values have significantly changed from one labelled set to another is the one discussed in Saerens et al. (2002, §3).

1.6   Quantification and Bias Mitigation

Quantification is inherently connected to the notion of *bias*, and to attempts at mitigating it. This is best explained by looking at the behaviour of the Classify and Count method in action. To this aim, let us consider IMDb, a dataset of movie reviews often used for evaluating binary sentiment quantification systems. The dataset consists of 50,000 documents and is perfectly balanced, i.e., there is an equal number of **Positive** and **Negative** reviews. Let us split the dataset in two equally-sized, perfectly balanced portions, one used for training purposes and another used for testing purposes. Let us use the training portion (containing 25,000 documents) to generate 9 random training samples of 5,000 documents each, at controlled class prevalence values. Specifically, let us sample random training sets $L_{10\%}$, $L_{20\%}$, ..., $L_{90\%}$ with a prevalence value for the class **Positive** of 10%, 20%, ..., 90%, respectively. Let us use each training set thus generated to train a classifier (in this case we use an SVM with a linear kernel), and we use each such classifier to implement a basic Classify & Count approach, thus generating a series of quantifiers that we denote by $CC_{10\%}$, $CC_{20\%}$, ..., $CC_{90\%}$. Let us do something similar in the test portion in order to generate test samples characterised by widely varying class prevalence values. In particular, let us use a finer-grained grid of prevalence values in order to generate test sets $U_{0\%}$, $U_{5\%}$, $U_{10\%}$, ..., $U_{100\%}$ of 500 documents each, and let us repeat this process 100 times in order to obtain more reliable results. Finally, we use all of our CC quantifiers to generate predictions for all the test samples. (Experimental protocols like the one we have described here are rather common in the quantification literature, and will be the subject of Section 3.4.) The results of this experiment are reported in Figure 1.5. These plots represent the estimated

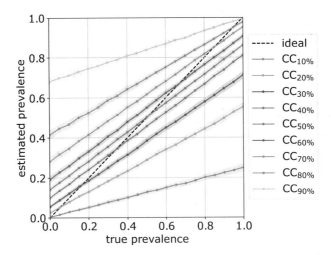

Fig. 1.5 Diagonal plot showing how CC delivers biased estimators of class prevalence values.

prevalence values along the y-axis and the true prevalence values along the x-axis; we show results averaged across the 100 repetitions, with colour bands representing standard deviation. Since IMDb is binary, we only report results for the Positive class. Such a plot is typically called a "diagonal plot", and will be more thoroughly discussed in Section 6.3.2.

The most important fact that emerges from this figure is that *Classify and Count generates biased estimators of class prevalence*, and is thus (as already anticipated in Section 1.2) a suboptimal quantification method: its prevalence estimates $\hat{p}_{U_\alpha}(y)$ for a class y are always intermediate between the true prevalence $\alpha = p_{U_\alpha}(y)$ in the unlabelled set U_α and the prevalence $\beta = p_{L_\beta}(y)$ in the labelled set L_β on which the classifier was trained (where $\alpha, \beta \in [0\%, 100\%]$), and are very often much closer to the latter than to the former. In other words, the factor that biases the class prevalence estimators is the class prevalence of the training set: in general, given sets of data L and U, Classify and Count does not seem to be able to predict class prevalence values for y much different from $p_L(y)$, even if the true class prevalence value $p_U(y)$ is faraway from this value. This trend is by no means specific to this dataset, and naturally arises in many different applicative contexts.

This should not surprise us, since standard learning mechanisms assume that the training set L and the unlabelled set U are IID, i.e., that $p_L(\mathbf{x}, y) = p_U(\mathbf{x}, y)$; as a result, the predictor learns from L not only the correlation between features and labels (i.e., $p(\mathbf{x}, y)$), but also the prevalence values of the labels (i.e., $p(y)$). An additional fact that emerges from Figure 1.5 is that the more the training set is imbalanced, the stronger this effect is; in fact, this effect is strongest in the extreme cases concerning the $L_{10\%}$ and $L_{90\%}$ datasets, while it is weakest in the perfectly balanced $L_{50\%}$ dataset.

As much recent research on the fairness and accountability of machine learning methods shows Mehrabi et al. (2019), sample selection bias may be a serious problem, in that it may propagate stereotypes and lead to incorrect decision-making. As an example, suppose our aim is to estimate the prevalence of class AfricanAmerican in an unlabelled set representing patients not covered by insurance (Elliott et al., 2009). If the prevalence of this class is .10 in the training set L, the estimate $\hat{p}(\text{AfricanAmerican})$ may be close to .10 even if the true prevalence of AfricanAmerican in U is, say, .50. This may lead to underestimating racial disparities in healthcare, misguided public health decisions, and diversion of precious resources.

The goal of "genuine" quantification methods (i.e., methods different from Classify and Count) is thus to eliminate, or at least mitigate, this bias; aside from Classify and Count, Figure 1.6 plots the results of CC along with two other methods for learning to quantify (in this case, all methods are trained on a perfectly balanced subset of 5,000 documents), and the fact that the curves corresponding to these other methods are closer to the diagonal line than the Classify and Count curve, shows that these other methods succeed, in varying degrees, in mitigating this bias. More on this in the sections to follow.

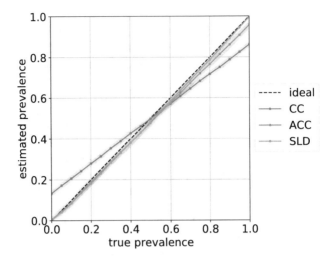

Fig. 1.6 Diagonal plot showing quantification methods that succeed in mitigating bias for the IMDb dataset.

1.7 Structure of This Book

In the above sections we have motivated why quantification is an interesting problem, why it should be addressed as a task of its own instead of as a byproduct of classification, how it is rooted in the fundamental problem of dataset shift, and how one of its goals is to mitigate the bias of which Classify and Count suffers.

The rest of this book is structured as follows.

Section 2 examines the applications of quantification. Special emphasis is given to the fields of human inquiry which are devoted to studying phenomena only at a collective level, such as market research, political science, the social sciences, ecological modelling, and epidemiology. However, we also pay attention to quantification as a means to improve, in scenarios characterised by distribution shift, the accuracy of *classification*, and this may in turn have an impact on many diverse fields and applications.

In Section 3 we turn our attention to the issue of how to experimentally evaluate quantification algorithms. A large part of this section is devoted, as should be expected, to discussing the various measures that have been proposed over the years for evaluating quantifiers. However, we also pay attention to the different experimental *protocols* that have been used in different works for carrying out the evaluation, protocols that differ essentially in terms of the stand they take towards relying or not on artificially generated samples.

Section 4 is devoted to presenting supervised learning methods for performing quantification, starting with "aggregative" methods (i.e., methods that involve the classification of individual items as a preliminary step) and ending with "non-aggregative" ones (i.e., methods that analyse the sample "holistically", without issuing individual classification decisions). In the course of this discussion, attention is paid both to methods that rely on "general-purpose" learners (i.e., ones that had originally been designed for tasks other than quantification) and to methods that are based on "special-purpose" learners (i.e., learners designed with quantification in mind).

In Section 5 we look at some advanced, "niche" topics, including quantification for ordinal codeframes, regression quantification, text quantification in cross-lingual settings, quantification for networked data, quantification for data streams, and others.

Section 6 takes a look back at the historical development of quantification as a task, and how (as many other tasks) it has witnessed independent contributions from researchers coming from different areas (machine learning, data mining, statistics, information retrieval), sometimes unaware of the developments that had gone on in other areas. This section also describes publicly available software packages and a brief tour of experimental results and visualisation tools to present them. Finally, we look at related tasks, spelling out the differences between them and learning to quantify. We conclude in Section 7, hinting at open problems and possible areas of further investigation.

Chapter 2
Applications of Quantification

Broadly speaking, there are two reasons why one might want to perform supervised prevalence estimation:

1. The first is that the estimated class prevalence values may serve the purpose of improving the accuracy or the fairness of a classifier. It is mostly machine learning researchers who have investigated supervised prevalence estimation from this angle, a task which they normally call *(class) prior estimation*. We look at this type of applications in Sections 2.1 and 2.2.
2. The second is that the estimated class prevalence values may be applicatively interesting by themselves, i.e., obtaining them is the final goal. It is mostly data mining / statistics / text mining researchers who have taken this angle, and it is here that the term *quantification* has been coined (by Forman, 2005) and adopted. We look at this type of applications in Sections 2.3 to 2.8.

2.1 Improving Classification Accuracy

The presence of dataset shift can damage the accuracy of a machine-learned classifier, because essentially all classifier training algorithms are based on the IID assumption (i.e., perform at their best when the training set L and the set U of unlabelled items are IID), an assumption which dataset shift invalidates. A particularly illuminating example of why distribution shift, of all the types of dataset shift that can occur, can make a classifier perform sub-optimally, is the Bayes optimal classifier, which is given by

$$h(\mathbf{x}) = \arg\max_{y} p(y|\mathbf{x})$$

$$= \arg\max_{y} \frac{p(\mathbf{x}|y)p(y)}{p(\mathbf{x})} \tag{2.1}$$

© The Author(s) 2023
A. Esuli et al., *Learning to Quantify*, The Information Retrieval Series 47,
https://doi.org/10.1007/978-3-031-20467-8_2

This equation shows that the posterior probabilities $p(y|\mathbf{x})$ generated by the classifier (and, in turn, the classification decision $\arg\max_y p(y|\mathbf{x})$) depend on the class prevalence values $p(y)$, which are estimated on L. In the presence of distribution shift between L and U this estimation will be inaccurate, and the quality of the posterior probabilities (and of the final decision) will be negatively influenced. For instance, if $p_U(y) > p_L(y)$, then $p(y|\mathbf{x})$ as deriving from Equation 2.1 will be smaller than it should be, and y will have a lower-than-ideal chance to be picked as the label for $\mathbf{x}$.

In order to improve the quality of both the posterior probabilities and the classification decisions generated by the classifier, we would need to use, in Equation 2.1, the value $p_U(y)$ in place of the value $p_L(y)$ that is normally used. Since $p_U(y)$ is unknown, one possibility is to use quantification methods to estimate it.

More precisely, what the use of quantification methods allows to do is to improve the *calibration* of the posterior probabilities. An intuition of what "calibrated probabilities" means is given by the following example. For instance, if only 10% of all the data items $\mathbf{x}$ for which $p(y|\mathbf{x}) = .5$ indeed belong to y, we can say that the classifier has overestimated the probability that these items belong to y, and that their posteriors are thus inaccurate; if this percentage is instead 90%, we can say that the classifier has underestimated this probability, again resulting in inaccurate posteriors. Indeed, we say (see e.g., Flach, 2017) that the posteriors $p(y|\mathbf{x})$, where the data items $\mathbf{x}$ belong to a sample σ, are (perfectly) *calibrated* (i.e., accurate) when, for all $a \in [0, 1]$, it holds that[1]

$$\frac{|\{\mathbf{x} \in \sigma \mid p(y|\mathbf{x}) = a, \Phi(\mathbf{x}) = y\}|}{|\{\mathbf{x} \in \sigma \mid p(y|\mathbf{x}) = a\}|} = a \tag{2.2}$$

Even assuming that our learner generates classifiers that tend to return well-calibrated probabilities, the classifier is calibrated for the training set L, which means that, in the presence of distribution shift, it cannot be calibrated for U too. The posteriors $p(y|\mathbf{x})$ can be re-calibrated (i.e., tuned on the unlabelled data) by multiplying them by $p_U(y)/p_L(y)$, but in order to do this, $p_U(y)$ needs to be estimated, which is where quantification comes into play. Well calibrated probabilities are important in a number of tasks, including (aside standard classification, as argued earlier in this section) (a) cost-sensitive classification (Elkan, 2001), (b) risk assessment and minimisation (as in credit scoring Hand and Henley, 1997 or in technology-assisted review Oard et al., 2018), and (c) ranking classes in terms of their suitability to a data item (Makris et al., 2007).

Most works that use quantification in order to improve classification accuracy do so, as explained above, by trying to improve the quality of the posterior probabilities;

---

[1] Perfect calibration is usually unattainable on any non-trivial dataset; however, calibration comes in degrees (and the quality of calibration can indeed be measured, via functions such as *calibration error*), so efforts can be made to obtain posteriors that are as close as possible to their perfectly calibrated counterparts.

this is the route that Alaíz-Rodríguez et al. (2011), Saerens et al. (2002), Vucetic and Obradovic (2001), Xue and Weiss (2009), and Zhang and Zhou (2010) follow. A different line of research is that of Balikas et al. (2015), who use quantification for optimising the parameters of the classifier in semi-supervised classification contexts in which there are not enough labelled validation data to optimise the parameters on.

2.1.1 Word Sense Disambiguation

Chan and Ng (2005, 2006) show that *Word Sense Disambiguation* (WSD) is a particularly interesting application context in which one might want, as discussed above, to improve the quality of the posterior probabilities with the goal of improving classification accuracy. WSD is the task of predicting, given a natural language sentence in which an ambiguous word occurs, which of the senses that this word has is the intended one. For each word it is assumed that there are a finite number of senses and that these senses are known in advance; as a result, this is a classification task, where the occurrence of the word is the item to classify and the senses of the word are the classes. As a result, word sense disambiguators are usually classifiers trained on corpora of sense-tagged texts.

However, these classifiers are often influenced by the sense priors of the corpora they have been trained on. For instance, assume that the word to disambiguate is bank, that one of its senses is that of a financial institution (as in the bank round the corner) and another of its senses is that of a hydraulic artefact (as in the banks of river Thames). Assume that such a classifier has been trained on a corpus L of financial texts; in this case the prevalence of the former sense will be much larger than that of the latter sense. Assume also that the trained classifier is used to disambiguate a set U of texts about hydraulic engineering; in this case many occurrences of word bank will have the latter sense but will be wrongly attributed to the former one, since the classifier is biased towards the financial sense of the word. One might thus want to recalibrate on the unlabelled set U the posterior probabilities of the different word senses, and the way Chan and Ng (2005, 2006) do so is by using quantification in the way discussed above.

Note that this is just an instance of the general process of adapting a classifier trained on a "source" domain to a different, "target" domain, a process known as *transfer learning* (Vilalta et al., 2011) which has countless applications. Since no part of the process described above is specific to word sense disambiguation, this suggests that quantification may play an important role in several other contexts in which transfer learning is used.

2.2 Fairness

2.2.1 Improving Fairness

Quantification can be used to improve not only the accuracy of a classifier h but also its *fairness*, i.e., its ability to avoid propagating prejudice, inequity, and partisanship. Biswas and Mukherjee (2021) use quantification in order to make sure that a classifier h does not promote discrimination with respect to a sensitive attribute, such as race or gender, and do so by introducing the notion of *Proportional Equality* (PEq). Suppose we mark a given attribute s as "sensitive" or "protected", i.e., we want to impose that it should not be a basis for discrimination. For the sake of exposition, let us consider binary sex as sensitive, with class values $c \in \mathcal{S} = \{♂, ♀\}$. It might well be that our training set L is sex-biased, i.e., for a certain class y it happens that $p_L(y|♂)$ (the prevalence of y in the set of male individuals belonging to L) is substantially different from its female counterpart $p_L(y|♀)$; for instance, if y corresponds to the class of **Engineers**, it might happen that $p_L(y|♂) \gg p_L(y|♀)$. It might also happen that our set U of unlabelled items does not have this bias, i.e., it does not hold that $p_U(y|♂) \gg p_U(y|♀)$. In this case, we would not want the bias in L to influence the way the data in U are classified.[2] This can be achieved by imposing proportional equality, i.e., imposing that

$$\text{PEq} = \left| \frac{p_U^h(\hat{y}|♂)}{p_U^h(\hat{y}|♀)} - \frac{p_U(y|♂)}{p_U(y|♀)} \right| \leq \epsilon \qquad (2.3)$$

where $p_U^h(\hat{y}|c)$ represents (see Table 1.1) the fraction of members of sample U that belong to c and to which classifier h has assigned class y. In other words, Equation 2.3 prescribes that the way the labels assigned by the classifier are distributed in U is "fair", i.e., reflects the way they are actually distributed in U. Of course, this latter distribution is unknown; the idea is thus to estimate it via quantification methods, and plug the resulting estimate of PEq into an optimisation procedure aimed at minimising it.

---

[2] A well-known example comes from machine translation. In the past, it was reported that services such as Google Translate or Microsoft Translator, when translating into English from gender-neutral languages such as Turkish (where, e.g., the personal pronoun "o" is used for males and females alike), tended to associate words such as "doctor" to male pronouns ("O bir doktor" → "He is a doctor"), while they tended to associate words such as "cook" to female pronouns ("O bir ahci" → "'She is a cook'), presumably due to gender bias present in the text corpora the translation service had been trained on. See (Emel Ince, *Do the footprints of stereotyping and gender bias follow us in online environments?*, 2018, https://www.capstan.be/do-the-footprints-of-stereotyping-and-gender-bias-follow-us-in-online-environments/, retrieved on Feb 28, 2020) for the full story.

2.2.2 Measuring Fairness

Quantification methods are also suited to "measuring (classifier) fairness under unawareness", i.e., providing estimates of the fairness of classifiers with respect to a sensitive attribute (e.g., race, sex) in situations where the values of the sensitive attribute are not available at classifier training and/or test time. This is a common setting in practice, due to several factors, including legislation on demographic data collection (Bogen et al., 2020), privacy-by-design standards, and a data minimisation ethos (Andrus et al., 2021). For this reason, the problem of measuring fairness under unawareness has become important for many practitioners interested in evaluating the differential impact of their classifiers across salient subpopulations, identified by sensitive attributes whose ground truth values are not known (Holstein et al., 2019).

Fabris et al. (2021) adapt quantification approaches to tackle the fairness-under-unawareness problem. For the sake of exposition, let us focus on "demographic parity" (Barocas et al., 2019; Calders and Verwer, 2010), a measure of classifier fairness focused on the difference in the values of "acceptance rate" (i.e., the fraction of data items that are assigned the class of interest) across different subpopulations (determined by sensitive attribute s) for a classifier $k : \mathcal{X} \rightarrow \mathcal{Y}$, issuing predictions $k(\mathbf{x})$ for a target variable (e.g., employability) across the data points (e.g., candidates).[3] Let us consider again, for the sake of exposition, binary sex as the sensitive attribute s, with class values $c \in \mathcal{S} = \{♀, ♂\}$, and employability as the target variable, with class values $y \in \mathcal{Y} = \{⊕, ⊖\}$ (where we assume that $⊕$ stands for "Hire" and $⊖$ stands for "Turn down"). The demographic disparity (DD) of classifier k with respect to sensitive attribute s is defined as

$$\mathrm{DD}(k, s, \sigma) = p_\sigma^k(\hat{⊕}|♀) - p_\sigma^k(\hat{⊕}|♂) \tag{2.4}$$

where

$$p_\sigma^k(\hat{⊕}|c) = p_\sigma^k(c|\hat{⊕}) \frac{p_\sigma^k(\hat{⊕})}{p_\sigma(c)} \tag{2.5}$$

is the acceptance rate for class c, and where Equation 2.5 is just an application of Bayes' theorem. Under this measure, classifiers are considered fair if their DD is close to zero, while extreme values of -1 or $+1$ indicate maximum unfairness, since the difference in acceptance rates across sensitive subpopulations is maximum. If $k(\mathbf{x})$ represents the employability of candidates, k as applied to σ is considered maximally fair under DD if the probability $p_\sigma^k(\hat{⊕})$ of being hired is the same for

---

[3] In this section, we let $k(\mathbf{x})$, instead of $h(\mathbf{x})$ as defined in Table 1.1, denote the hard classifier issuing predictions in $\mathcal{Y}$, since here sensitive attributes in $\mathcal{S}$ are the target of quantification. In other words, we reserve the notation $h(\mathbf{x})$ for a hard classifier issuing predictions in the same domain of the quantification task.

males and females, which would mean that DD$(k, s, \sigma) = 0$. Due to the difficulties in demographic data procurement outlined above, the values for sensitive attribute s are often unknown at classifier training and/or test time. The value of $p_\sigma^k(\hat{\oplus}|c)$ can be computed if we have reliable estimates of groupwise prevalence values $p_\sigma^k(c|\hat{\oplus})$ and $p_\sigma^k(c|\hat{\ominus})$, since Equation 2.5 can be re-written as

$$p_\sigma^k(\hat{\oplus}|c) = p_\sigma^k(c|\hat{\oplus}) \frac{p_\sigma^k(\hat{\oplus})}{p_\sigma^k(\hat{\oplus}) \cdot p_\sigma^k(c|\hat{\oplus}) + p_\sigma^k(\hat{\ominus}) \cdot p_\sigma^k(c|\hat{\ominus})} \quad (2.6)$$

In other words, since $p_\sigma^k(\hat{\oplus})$ and $p_\sigma^k(\hat{\ominus})$ are available, DD(k, s, σ) (Equation 2.4) can be readily estimated by leveraging quantification methods that provide estimates $p_\sigma^k(c|\hat{\oplus})$ and $p_\sigma^k(c|\hat{\ominus})$. A necessary requirement for this is the availability of a (possibly small) auxiliary annotated dataset L in which the values of the sensitive attribute are the labels. This dataset is to be used for training the quantifier that must be applied to σ, and may derive from voluntary data disclosures, surveys, or other targeted efforts.

However, because of their nature, these datasets are likely to suffer from selection bias, and unlikely to be fully representative of the deployment conditions. Fabris et al. (2021) show that quantification methods are particularly suited to tackle the fairness-under-unawareness problem, as they are robust to the inevitable distribution shift that derives e.g., from selection bias. Moreover, the authors show that quantification methods can effectively decouple the (desirable) objective of measuring classifier fairness from the (undesirable) side effect of allowing the inference of the sensitive attribute values of individuals, thus reducing the potential for model misuse at the individual level (e.g., profiling).

2.3 Sentiment Analysis

Sentiment classification, the task of classifying a piece of text about a certain object as expressing a positive, neutral, or negative sentiment toward that object, has become a ubiquitous enabling technology, with applications in many fields, including financial news analysis, brand positioning and reputation management, stock market prediction, customer relationships management, and others.

Of interest to us is the fact that, while in some applications the sentiment of a specific individual is of interest, in other cases the application requirements only involve assessing the sentiment of a certain population (Esuli and Sebastiani, 2010b); for instance, while customer relationship management is typically an application of the former type, brand positioning is usually concerned with collective sentiment only. In particular, Gao and Sebastiani (2016) observe that most endeavours having to do with sentiment classification in Twitter are really about *sentiment quantification*, since hardly anybody who sets out to classify tweets by sentiment, is interested in the sentiment expressed in specific, individual tweets.

Collective sentiment is often an object of study in the social and political sciences, as well as in market research; much of what is discussed in the next two sections, which are devoted to applications in these disciplines, touches on sentiment-related issues too.

2.4 Social and Political Sciences

The social and political sciences are disciplines in which individual cases hardly matter, and where the interest is instead on phenomena that require analysis at the aggregate level.

One of the many examples of this (and of the rising field of *computational social science*) is illustrated in Figure 2.1 (from Dodds et al., 2011). Here, the authors set out to study the temporal patterns of happiness in the population of Twitter users. Essentially, what the authors do is to engage in some type of sentiment classification (Happy vs. Unhappy) that detects whether a certain tweet denotes happiness or unhappiness, bin the results according to the time and date the corresponding tweets were issued, and plot the Happy and Unhappy relative frequencies of the corresponding bins on a temporal scale. This endeavour has two characteristics that are of interest to us. The first is that the objects of interest (the tweets) are unlabelled, i.e., it is not known (and it is not possible to deterministically determine) whether they are representative of the class Happy or not. The second is that the authors are not interested in individual tweets, but in the big picture,

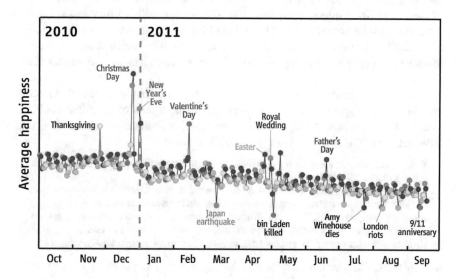

Fig. 2.1 Temporal patterns of happiness as resulting from a Twitter study (from (Dodds et al., 2011)).

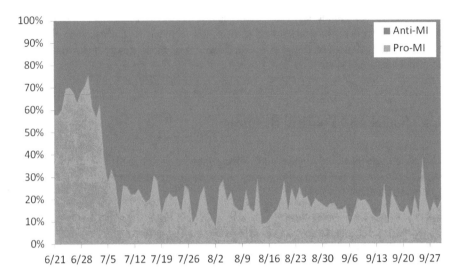

Fig. 2.2 Temporal trend in the proportions of tweets supporting or opposing military intervention in Egypt during the "Arab spring" in summer 2013 (from Borge-Holthoefer et al., 2015).

i.e., in the proportions (at different time points) of tweets that belong or do not belong to class **Happy**. This is thus a case where quantification (actually: sentiment quantification) techniques could have been applied. Yet a further example is reported in Figure 2.2 (from Borge-Holthoefer et al., 2015), where the authors mine the Twittersphere in order to determine (among other things) the prevalence values of the pro-military-intervention stance vs. the against-military-intervention stance concerning the summer 2013 "Arab spring" in Egypt. Again, we have a combination of unlabelled data and interest at the aggregate level only, which would have made this research suitable for the application of *stance quantification* (see Walker et al., 2012).

Concerning the fact that social scientists are interested in phenomena that require analysis at the aggregate level, we simply echo the words of Hopkins and King (2010), who have been the first to use a non-trivial quantification method (i.e., one different from "classify and count") for political analysis reasons:

> When social scientists use formal content analysis, it is typically to make generalisations using document category proportions. (. . .)
> Policy-makers or computer scientists may be interested in finding the needle in the haystack (. . .) but social scientists are more commonly interested in characterising the haystack. (. . .) Although computer scientists have methods for automated content analysis, most are optimised to classify individual documents, whereas social scientists instead want generalisations about the population of documents, such as the proportion in a given category.

In their work, these authors use the ReadMe algorithm (that we will analyse in detail in Section 4.4.1) with the aim of estimating the prevalence of different political candidates in bloggers' preferences through an analysis of their blog posts. In other works in a similar spirit, researchers have variously tried to estimate the distribution of press releases related to legislators' credit claiming efforts (Grimmer et al., 2012), to estimate the prevalence of different types of censored news in Chinese media (King et al., 2013), and to estimate the distribution of citizens' political preferences by performing sentiment analysis on tweets (Ceron et al., 2014).

It has to be noted that, while the above-mentioned works use a quantification method other than Classify and Count, the vast majority of works in the social and political sciences that make use of supervised learning use Classify and Count (see Mandel et al., 2012 for an example), no doubt due to a lack of awareness of the sub-optimality of this strategy, of the existence of better alternatives, and of the very existence of quantification as a task. This state of affairs is not limited to the social and political sciences, though, and cuts across all the disciplines to which quantification has been applied (or could be applied), and which will be mentioned in the next sections.

2.5 Market Research

The goal of market research is to obtain information concerning the desires and needs of actual or potential customers of products or services. This information is usually collected through surveys, conducted by a survey specialist and involving a number of respondents. Conducting a survey usually involves a questionnaire, i.e., a list of questions which respondents are asked to answer. The majority of questions to be found in questionnaires are of the "closed" type, where the respondent is required to tick one of a predefined set of answers. Open (a.k.a. "open-ended") questions instead involve returning a textual answer. When computing the results of the survey, in order to manage open questions the survey specialist first defines a set of classes of interest for the given application (e.g., HatesSitComs, WantsMoreSoaps, etc., for a survey run on behalf of a TV network), and then classifies (either manually or via a machine-learned classifier) each answer based on its textual content. The results of the survey are then obtained by checking how many respondents' answers have been attributed which class. Quite obviously, the focus on "how many" (as opposed to "which") in the previous sentence indicates that the survey specialist needs, rather than a classifier, a quantifier of open-ended answers.

The use of non-trivial methods for performing quantification for open-ended answers in market research has been proposed only very recently (Sebastiani, 2018). Unsurprisingly, previous literature (see Esuli and Sebastiani, 2010a for an example) just reports uses of Classify and Count, for pretty much the same reasons as described in Section 2.4.

2.6 Epidemiology

Epidemiology is a discipline with traits analogous to the social and political sciences, since the objects of study are individuals but the quantities of interest are only indicators at the aggregate level. Epidemiologists try to obtain estimates of disease prevalence values across different geographical regions, time periods, age groups, or gender. (See the example in Figure 2.3.) These prevalence estimates are important in assessing the spread of infectious diseases, in assessing the impact of toxic environmental conditions, in planning and allocating health services, and in measuring health risks.

One way quantification may be applied to epidemiology is in establishing disease prevalence by analysing, via text quantification techniques, clinical reports of a textual nature. One such example is reported in Baccianella et al. (2013), where the authors quantify (using Classify and Count) over a dataset of radiology reports. Applications such as this are difficult, especially when it comes to rare diseases. In these cases, as already mentioned in Bullet 2 of Section 1.5, when creating the training set the classes that represent rare illnesses need to be oversampled, in order to improve the accuracy of the predictor; the distribution of these classes in the training data may thus be *very* different from the distribution in the unlabelled data, thus generating situations of extreme distribution shift.

Estimated TB incidence rates, 2019

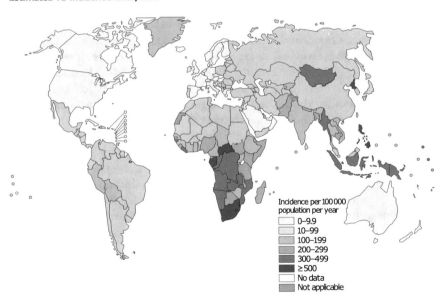

Fig. 2.3 The prevalence of tubercolosis in 2019, expressed as number of cases per 100,000 inhabitants (from the *Global Tubercolosis Report*, Geneva: World Health Organization, 2020).

Another (perhaps more peculiar) application of quantification to epidemiology that has been reported is the estimation of the prevalence of various causes of death via "verbal autopsies" (King and Lu, 2008). A verbal autopsy is a textual description of the symptoms that a deceased person exhibited before dying; this description may be obtained from family members or other caretakers. Such a description may be used in order to later establish the causes of death of the deceased in situations in which a doctor entitled to certify these causes is not available; example such scenarios are remote villages in developing countries, or in areas faraway from hospitals. Using a verbal autopsy in such a way can be framed as a supervised classification problem, using classification schemes where all known causes of death are organised in a taxonomy: a text classifier classifies the verbal description of symptoms (which can be represented as a vector of features) and assigns it the classes in the taxonomy that befit this description. In order to train such a classifier, training data may be obtained at hospitals, since for patients that have deceased in a hospital both a description of symptoms obtained from nurses and doctors *and* the causes of death as certified by a doctor, are typically available. When these causes of death are needed for reasons other than epidemiological ones, a classifier is the desirable tool; for the needs of epidemiology, instead, a quantifier is the most adequate one. Note that in typical scenarios of the above type, distribution shift is at play, for various reasons. One reason is the same as mentioned for the analysis of clinical reports, i.e., rare diseases requiring oversampling for creating the training data. A second reason is that the training set may have been collected by merging different datasets from hospitals in different geographical areas. Yet another reason is that local environmental conditions in the application scenario (e.g., a nearby toxic industrial plant) may make these conditions irreproducible at training time. One application of quantification to establishing the prevalence of causes of death for epidemiological purposes is reported in King and Lu (2008) and King et al. (2010), where the ReadMe quantification method (to be discussed in Section 4.4.1) is used.

A further type of application involves analysing social media posts in order to obtain indicators and trends related to public health. As reported by Daughton and Paul (2019), "classifying and counting" posts that allow to infer specific health-related characteristics of the person who has posted them, has been widely used, for applications ranging from influenza surveillance to measuring attitudes towards vaccination. However, only Daughton and Paul (2019) themselves, in yet another influenza surveillance application, have approached these problems by using quantification methods other than the trivial Classify and Count.

2.7 Ecological Modelling

When attempting to characterise ecosystems in order to allow their management and preservation, ecologists often need to assess the distribution of certain species across land and sea. When individual living beings cannot be characterised with certainty as belonging to a certain species or not, classification (carried out either manually or via a trained classifier) needs to be employed. However, ecologists are often interested in characterising not individual living beings, but entire populations of them; this is where quantification comes into play. A work in this direction is the one by González et al. (2017), which applies quantification technology in order to estimate the distribution of various plankton species in images of sea water samples.

A similar situation arises with land cover (LC) mapping. Given an aerial (e.g., satellite) image, LC mapping has to do with characterising how much of the territory represented in the image is covered in, e.g., forest, cultivated land, water, urban areas, etc. In order to do so, each pixel of the image is classified by an automated classifier as belonging to one of the above LC types. However, since we are only interested in predicting *how much* of the territory represented in the image is covered in a certain LC type, we may replace a classifier with a quantifier. This is the approach taken by Latinne et al. (2001), who (using the method that we will describe in Section 4.2.9) perform LC mapping on Landsat satellite images.

A work in the same spirit is Beijbom et al. (2015), where quantification is used to monitor the world's coral reefs by performing quantification on underwater images of seabed cover. Here, the objective is to estimate the percentages of seabed covered by each of 32 different species (see Figure 2.4), and distribution shift is caused by the fact that the location where the training images have been acquired is typically different, and thus exhibits a different distribution of species, from the locations where the unlabelled images are obtained (see Figure 1.2).

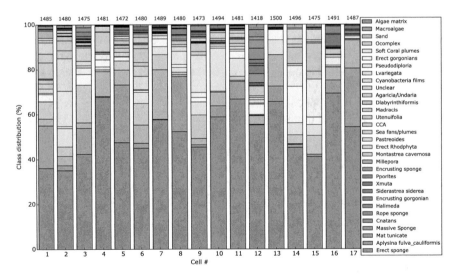

Fig. 2.4 Class prevalence of each of 32 living species in seabed cover as estimated via quantification technology (from Beijbom et al., 2015); the different columns represent different samples on which quantification has been performed.

2.8   Resource Allocation

Companies need to carefully plan how to allocate and distribute human resources to specific departments of the company, and must anticipate the needs of these departments in order not to be caught off-guard when the amount of work for a certain department spikes due to unusual circumstances.

In a series of papers, Forman (2005, 2006, 2008) describes an application of supervised prevalence estimation to resource allocation within a company. His work consists of automatically classifying the transcripts of phone calls received at the customer support department of a large IT company, where the classes are the different types of issues that such a customer support department is routinely asked to solve. Since the goal is to detect which issues are more prevalent, and thus need more personnel to be allocated to them, he proposes to use text quantification (instead of classification) technology. A correct estimation of the prevalence of the different issues not only allows a more adequate allocation of human resources: if performed systematically it allows to identify increasingly prevalent issues before they get out of control, to monitor if the resource allocation thus performed has been effective, and to focus product re-engineering / redesign efforts on the areas where this effort is most needed.

Chapter 3
Evaluation of Quantification Algorithms

As all other supervised learning algorithms, quantification algorithms must be subjected to a thorough experimental evaluation, and a pillar of this evaluation is the mathematical measure to be used. Sections 3.1 to 3.3 thus review the main evaluation measures for quantification that have been proposed for the various problems discussed in Section 1.4.

As already hinted in Section 1.1, quantification may be seen as generating a predicted distribution $\hat{p}$ over $\mathcal{Y}$ that approximates a true distribution p over $\mathcal{Y}$. Evaluating quantification thus means measuring how well $\hat{p}$ fits p. We will thus be concerned with discussing functions that attempt to measure this goodness-of-fit; we hereafter use the notation $D(p, \hat{p})$ to indicate such a function.

In this book we assume that the evaluation measures we are concerned with are measures of quantification error, and not of quantification accuracy. The reason for this is that most, if not all, the evaluation measures for quantification that have been used so far are indeed measures of error, so it would be slightly unnatural to frame everything in terms of quantification accuracy. Since any measure of accuracy can be turned into a measure of error (typically: by taking its negation), this is an inessential factor anyway.

A further problem in evaluating quantification is how we should choose the dataset and the samples on which to carry out this evaluation, in order for them to be representative of the scenarios encountered in real-world applications. This is a particularly thorny issue, since available datasets might not exhibit the type of shift, or the amount of shift, that we might want our quantifiers to be robust to. Section 3.4 thus discusses the different experimental protocols that have been proposed in the literature in order to address this problem.

© The Author(s) 2023
A. Esuli et al., *Learning to Quantify*, The Information Retrieval Series 47,
https://doi.org/10.1007/978-3-031-20467-8_3

3.1 Measures for Evaluating SLQ, BQ, and MLQ

In this section we will discuss a number of evaluation measures that have been proposed in the literature for evaluating single-label quantification.[1] As mentioned in Section 1.4, these measures can also be used in order to evaluate binary quantification and multi-label quantification, since BQ is a special case of SLQ, and since evaluating the error of a multi-label quantifier can be done by evaluating its BQ error for each $y \in \mathcal{Y}$. Many of the measures that we discuss here were originally proposed for BQ, but can be easily extended to deal with SLQ in general; we present them in their SLQ form, even when they were originally proposed in BQ form.

Essentially all evaluation measures that have been proposed in the quantification literature are *divergences*. Formally, a divergence D is a measure of how a predicted distribution $\hat{p}$ "diverges" (i.e., differs) from the true distribution p, and is such that (1) $D(p, \hat{p}) = 0$ if and only if $p = \hat{p}$, and (2) $D(p, \hat{p}) > 0$ for all $\hat{p} \neq p$. As an aside, note that two distributions p and $\hat{p}$ over $\mathcal{Y}$ are essentially two nonnegative-valued, length-normalised vectors of dimensionality $|\mathcal{Y}|$. The literature on the evaluation measures for quantification thus obviously intersects the literature on functions for computing the similarity of two vectors.

We here need to stress a key difference between measures of classification accuracy and measures of quantification accuracy (or error). The objects of classification are individual unlabelled items, and all measures of classification accuracy (e.g., F_1) are defined with respect to a test *set* of such objects. The objects of quantification, instead, are samples, and all the measures of quantification error we will discuss in this book are defined on a *single* such sample (i.e., they measure how well the true distribution of the classes *across this individual sample* is approximated by the predicted distribution of the classes across the same sample). Since every evaluation is worthless if carried out on a single object, it is clear that quantification systems need to be evaluated on *sets* of samples. This means that every measure that we are going to discuss needs first to be evaluated on each sample, and then its global score across the test set (i.e., the set of samples on which testing is carried out) needs to be computed. This global score may be computed via any measure of central tendency, e.g., via an average, or a median, or other.

---

[1] The measures discussed in this section are just the most frequently used ones, and are by no means an exhaustive list. E.g., other functions that have occasionally been used as evaluation measures for quantification are the *Pearson Divergence* (Ceron et al., 2016) and the *Discordance Ratio* (Levin and Roitman, 2017).

3.1.1 Properties of Evaluation Measures for SLQ, BQ, and MLQ

The most thorough published study of evaluation measures for SLQ is probably (Sebastiani, 2020). This paper defines a number of interesting formal properties that an evaluation measure for SLQ may or may not enjoy, discusses if (and when) each of these properties is desirable, and analyses whether the evaluation measures that have been used in the quantification literature enjoy them or not; this process is typical of the so-called *axiomatic approach* to "evaluating evaluation", i.e., to the study of evaluation measures (Busin and Mizzaro, 2013), an approach that has also been applied to other tasks such as classification and clustering. A significant result of this paper is that no existing evaluation measure for SLQ satisfies all the properties identified as desirable; still, some evaluation measures are proven to be "less inadequate" than others. We here briefly discuss four main such properties, mostly by way of examples. Sebastiani (2020) discusses still other properties, but these are satisfied by all the evaluation measures for quantification proposed in the literature, and as such are less interesting.

The first property we discuss here is *Maximum* (**MAX**). Basically, an evaluation measure for SLQ that enjoys **MAX** is one whose values are upper-bounded by a value $\beta > 0$, which is the same for all $\mathcal{Y}$ and for all p, and which is such that $D(p, \hat{p}) = \beta$ for at least one predicted distribution $\hat{p}$, called the *perverse* (i.e., worst possible) *estimator*. An evaluation measure that enjoys **MAX** is such that its range (or better: its image) is independent of the problem setting, and this allows to easily judge whether a given value of D means high or low quantification error; in other words, should this range depend on $\mathcal{Y}$, or on its cardinality, or on the true distribution p, we would not be able to easily interpret the meaning of a given value of D. An additional, possibly even more important reason for requiring this range to be independent of the problem setting is that, in order to test a given quantification method, the measure needs (as noted above) to be evaluated on a set of n test samples $\sigma_1, \ldots, \sigma_n$ (each characterised by its own true distribution), and a measure of central tendency across the n resulting values then needs to be computed. If, for these n samples, the measure ranges on n different intervals, this measure of central tendency will return unreliable results, since the results obtained on the samples characterised by the wider such intervals will exert a higher influence on the resulting value.

The second property is *Impartiality* (**IMP**). In essence, an evaluation measure D that enjoys **IMP** equally penalises the underestimation of a true prevalence $p(y)$ by an amount a (i.e., returning $\hat{p}(y) = p(y) - a$) or its overestimation by the same amount a (i.e., returning $\hat{p}(y) = p(y) + a$). This makes sense, because underestimation and overestimation should be considered equally undesirable, unless there is a specific reason (i.e., application need) for not doing so; in the latter case, the measure we choose should make its bias explicit, i.e., include a tunable parameter (similar in spirit to the β parameter of F_β) that allows specifying *how much* underestimation should be penalised more/less than overestimation.

The third property is *Relativity* (**REL**). In a nutshell, an evaluation measure that satisfies **REL** sanctions that an error of absolute magnitude a (i.e., the error made when $\hat{p}(y) = p(y) \pm a$) is more serious when the true class prevalence is smaller. In some applications of quantification **REL** is indeed desirable. Consider, as an example, the case in which the prevalence $p(y)$ of a certain cause of death y in a population has to be estimated, as discussed in Section 2.6, from "verbal autopsies". In this case, the evaluation measure should arguably enjoy **REL**; in fact, predicting $\hat{p}(y) = 0.0101$ when $p(y) = 0.0001$ is a much more serious mistake than predicting $\hat{p}(y) = 0.1100$ when $p(y) = 0.1000$, since in the former case a very rare cause of death is overestimated by two orders of magnitude, while the same is not true in the latter case.

The fourth property is *Absoluteness* (**ABS**), and is the opposite of **REL**. Basically, an evaluation measure that satisfies **ABS** sanctions that an error of absolute magnitude a should be penalised independently of the value of the true class prevalence. Of course, an evaluation measure cannot enjoy **REL** and **ABS** at the same time; however, while there are applications that require **REL**, other applications require **ABS**. Consider an example in which we want to predict the prevalence of the NoShow class among the passengers booked on a flight with actual capacity n (so that the airline can "overbook" additional $\hat{p}(\text{NoShow}) \times n$ seats). Here the evaluation measure should enjoy **ABS**, since returning $\hat{p}(\text{NoShow}) = 0.05$ when $p(\text{NoShow}) = 0.10$ or returning $\hat{p}(\text{NoShow}) = 0.15$ when $p(\text{NoShow}) = 0.20$ brings about the same cost to the airline (i.e., that $0.05 \times n$ seats will remain empty).

Note that, while **REL** and **ABS** are mutually exclusive, they do not cover the entire space of possibilities, i.e., there can be measures that enjoy neither **REL** nor **ABS**. One such measure is *cosine distance*, which, as it can be shown, considers an error of absolute magnitude a *less* serious when the true class prevalence is smaller.[2]

We will frame the discussion of evaluation measures for SLQ in terms of these four properties; for each such property and for each measure discussed in the following sections, Sebastiani (2020) presents proofs of whether the measure enjoys or does not enjoy the property.[3]

---

[2] Cosine distance will not be discussed any further in this book, because it has never been proposed or used as an evaluation measure for SLQ, and because a measure that enjoys neither **REL** not **ABS** is arguably of little use in any application of quantification.

[3] Note that there are several other properties that the literature on divergence functions and distance functions discusses, and that we do not consider here because we do not deem them interesting when it comes to evaluating quantification. For instance, one of them is *symmetry*, i.e., the property that states that for any two distributions p' and p'' it holds that $D(p', p'') = D(p'', p')$; in evaluating quantification we are not interested in symmetry, because our two distributions are not just any two distributions, but are always a true distribution and a predicted distribution, and switching their roles is not interesting.

3.1.2 Bias

Bias (B), defined as

$$B(y) = \hat{p}(y) - p(y) \tag{3.1}$$

is technically not an evaluation measure for quantification as we have defined it before, since it does not apply to an entire distribution p but only to a specific label y. Even when using it in a binary setting, one thus needs to specify which of the two classes it is applied to. It is a fairly simplistic measure, and we cover it only since it has been used in several papers on quantification (e.g., Forman, 2005, 2006; Tang et al., 2010).

A positive B score indicates that the prevalence of y has been overestimated, while a negative score indicates that it has been underestimated. If used as an evaluation measure for quantification, an obvious problem with B is that averaging the scores across different classes brings about unintuitive results, since the positive bias for one class and the negative bias for another class cancel each other out. The same problem occurs when sticking to the same class but averaging across different samples.

As a result, this measure can at most be used to determine if a method has a tendency to underestimate or overestimate the prevalence of a specific class (typically: the minority class) in BQ, and not as an evaluation measure for general use.

3.1.3 Absolute Error and its Variants

Absolute Error (AE), defined as

$$AE(p, \hat{p}) = \frac{1}{|\mathcal{Y}|} \sum_{y \in \mathcal{Y}} |\hat{p}(y) - p(y)| \tag{3.2}$$

is similar, but enforces the notion that positive and negative bias are equally undesirable. As a result, averaging it across several classes, or several samples, is not problematic.

As shown in Sebastiani (2020), AE enjoys **IMP** and **ABS** but does not enjoy **MAX** (and, since it enjoys **ABS**, does not enjoy **REL** either), since AE ranges between 0 (best) and

$$z_{AE} = \frac{2(1 - \min_{y \in \mathcal{Y}} p(y))}{|\mathcal{Y}|} \tag{3.3}$$

(worst), i.e., its range depends on the true distribution p and on the cardinality of $\mathcal{Y}$.

If viewed as a generic function of dissimilarity between vectors (and not just probability distributions), AE is nothing else than the well-known "city-block distance" normalised by the number of classes. Note that AE often goes by the name of *Mean* Absolute Error; for simplicity, for this and the other measures we discuss in the rest of this book we will omit the qualification "Mean", since every measure mediates across the class-specific values in its own way. Some recent papers Beijbom et al. (2015); González et al. (2017) that tackle quantification in the context of ecological modelling discuss or use, as an evaluation measure for quantification, *Bray-Curtis dissimilarity* (BCD), a measure popular in ecology for measuring the dissimilarity of two samples. However, when used to measure the dissimilarity of two probability distributions, BCD defaults to AE; as a result we will not analyse BCD any further.

Normalised Absolute Error (NAE), defined as

$$\text{NAE}(p, \hat{p}) = \frac{\text{AE}(p, \hat{p})}{z_{\text{AE}}} = \frac{\sum_{y \in \mathcal{Y}} |\hat{p}(y) - p(y)|}{2(1 - \min_{y \in \mathcal{Y}} p(y))} \tag{3.4}$$

is a version of AE that always ranges between 0 (best) and 1 (worst), and thus enjoys **MAX**. However, NAE does not enjoy **ABS** (while at the same time not enjoying **REL** either).

A slight variant of absolute error is *Squared Error* (SE), defined as

$$\text{SE}(p, \hat{p}) = \frac{1}{|\mathcal{Y}|} \sum_{y \in \mathcal{Y}} (\hat{p}(y) - p(y))^2 \tag{3.5}$$

It obviously shares the same pros and cons of AE, and we will not discuss it any further.

For AE and for all the other evaluation measures for quantification discussed in this book, Table 3.1 (reproduced from Sebastiani (2020)) lists the papers where the measure has been proposed and those which have subsequently used it for evaluation purposes.

3.1.4 Relative Absolute Error and its Variants

Relative Absolute Error (RAE), defined as

$$\text{RAE}(p, \hat{p}) = \frac{1}{|\mathcal{Y}|} \sum_{y \in \mathcal{Y}} \frac{|\hat{p}(y) - p(y)|}{p(y)} \tag{3.6}$$

is a refinement of AE that enforces **REL** by making AE relative to true class prevalence. RAE enjoys **IMP** and **REL** but does not enjoy **MAX** and (obviously)

Table 3.1 Research works about quantification where the evaluation measures for quantification discussed in this book have been first proposed (★) and later used (✓).

	AE	NAE	RAE	NRAE	SE	DR	KLD	NKLD	PD
Saerens et al. (2002)	★								
Forman (2005)	✓						★		
Forman (2006)	✓						✓		
Forman (2008)	✓						✓		
Tang et al. (2010)	✓						✓		
Bella et al. (2010)	✓				★				
González-Castro et al. (2010)	✓		★						
Zhang and Zhou (2010)	✓								
Alaíz-Rodríguez et al. (2011)	✓		✓						
Milli et al. (2013)							✓		
Barranquero et al. (2013)	✓								
González-Castro et al. (2013)	✓		✓						
Esuli and Sebastiani (2014)	✓	★	✓	★			✓	★	
du Plessis and Sugiyama (2014)					✓				
Esuli and Sebastiani (2015)			✓				✓		
Gao and Sebastiani (2015)	✓	✓	✓	✓			✓	✓	
Barranquero et al. (2015)	✓						✓		
Beijbom et al. (2015)	✓								
Milli et al. (2015)							✓		
Gao and Sebastiani (2016)	✓	✓	✓	✓			✓	✓	
Ceron et al. (2016)	✓								★
Kar et al. (2016)							✓		
Nakov et al. (2016)							✓		
González et al. (2017)	✓								
du Plessis et al. (2017)					✓				
Levin and Roitman (2017)						★			
Pérez-Gállego et al. (2017)	✓				✓				
Tasche (2017)			✓						
Nakov et al. (2017)	✓		✓				✓		
Maletzke et al. (2017)	✓	✓		✓					
Maletzke et al. (2018)					✓				
Esuli et al. (2018)	✓		✓				✓		
Card and Smith (2018)	✓								
Moreira dos Reis et al. (2018a)	✓								
Moreira dos Reis et al. (2018b)	✓				✓				

(continued)

Table 3.1 (continued)

	AE	NAE	RAE	NRAE	SE	DR	KLD	NKLD	PD
Sanya et al. (2018)							✓		
Keith and O'Connor (2018)	✓								
Fernandes Vaz et al. (2019)					✓				
Pérez-Gállego et al. (2019)	✓				✓				
Spence et al. (2019)	✓								
Hassan et al. (2020)	✓								
Esuli et al. (2020)	✓		✓				✓		
Qi et al. (2020)	✓		✓				✓		
Alexandari et al. (2020)					✓				
Fabris et al. (2021)	✓				✓				
Biswas and Mukherjee (2021)	✓								
Moreo and Sebastiani (2021)	✓		✓						
Moreo et al. (2021a)	✓								
Schumacher et al. (2021)	✓							✓	
Moreo and Sebastiani (2022)	✓		✓						
Jerzak et al. (2022)	✓								

ABS. It does not enjoy **MAX** because it ranges between 0 (best) and

$$z_{\text{RAE}} = \frac{|\mathcal{Y}| - 1 + \dfrac{1 - \min\limits_{y \in \mathcal{Y}} p(y)}{\min\limits_{y \in \mathcal{Y}} p(y)}}{|\mathcal{Y}|} \tag{3.7}$$

(worst), i.e., its range depends on the true distribution p and on the cardinality of $\mathcal{Y}$. *Normalised Relative Absolute Error* (NRAE), a version of RAE that ranges between 0 (best) and 1 (worst), can thus be obtained as

$$\text{NRAE}(p, \hat{p}) = \frac{\text{RAE}(p, \hat{p})}{z_{\text{RAE}}} = \frac{\sum\limits_{y \in \mathcal{Y}} \dfrac{|\hat{p}(y) - p(y)|}{p(y)}}{|\mathcal{Y}| - 1 + \dfrac{1 - \min\limits_{y \in \mathcal{Y}} p(y)}{\min\limits_{y \in \mathcal{Y}} p(y)}} \tag{3.8}$$

However, it can be shown that NRAE does not enjoy **REL** (and does not enjoy **ABS** either), so its name "Normalised Relative Absolute Error" is somehow a misnomer.

Note that both RAE and NRAE may be undefined due to the presence of zero denominators. To solve this problem, in computing RAE and NRAE we can smooth both $p(y)$ and $\hat{p}(y)$ via additive smoothing, i.e., we take

$$\underline{p}(y) = \frac{\epsilon + p(y)}{\epsilon|\mathcal{Y}| + \sum_{y\in\mathcal{Y}} p(y)} \tag{3.9}$$

where $\underline{p}(y)$ denotes the smoothed version of $p(y)$ and the denominator is just a normalising factor (same for the $\hat{p}(y)$'s); the quantity $\epsilon = \frac{1}{2|U|}$ is often used as a smoothing factor. The smoothed versions of $p(y)$ and $\hat{p}(y)$ are then used in place of their original non-smoothed versions in Equations 3.6 and 3.8; as a result, RAE and NRAE are always defined. The same method will also be used for all other measures that may incur in the problem of zero denominators (see e.g., Equation 3.10), and that we will encounter in the next sections.

3.1.5 Kullback-Leibler Divergence and its Variants

Forman (2005) proposes to evaluate SLQ by means of *normalised cross-entropy*, better known as *Kullback-Leibler Divergence* (KLD). KLD, defined as

$$\mathrm{KLD}(p, \hat{p}) = \sum_{y\in\mathcal{Y}} p(y) \log \frac{p(y)}{\hat{p}(y)} \tag{3.10}$$

ranges between 0 (best) and $+\infty$ (worst). KLD is widely used as an evaluation measure for SLQ, and it has also been adopted as the official evaluation measure of the only quantification-related shared task that has been organised so far, Subtask D "Tweet Quantification on a 2-point Scale" of SemEval-2016 and SemEval-2017 "Task 4: Sentiment Analysis in Twitter" (Nakov et al., 2016, 2017).

The fact that KLD is not upper-bounded means that it does not satisfy **MAX**.[4] *Normalised Kullback-Leibler Divergence* (NKLD), defined as

$$\mathrm{NKLD}(p, \hat{p}) = 2\frac{e^{\mathrm{KLD}(p,\hat{p})}}{e^{\mathrm{KLD}(p,\hat{p})} + 1} - 1 \tag{3.11}$$

is a variant of KLD that does enjoy **MAX**, since it ranges between 0 (best) and 1 (worst). Unfortunately, as shown in Sebastiani (2020), both KLD and NKLD enjoy

---

[4] Actually, the fact that smoothing is used makes KLD upper-bounded, but by a factor that depends on both p and $\mathcal{Y}$, which means that KLD does not satisfy **MAX** anyway. See Sebastiani (2020, §4.7) for details.

none of **IMP**, **REL** and **ABS**, which makes their use as evaluation measures for quantification questionable.

A further problem of KLD and NKLD is that they score low in terms of understandability, i.e., look esoteric to the mathematically uninitiated, at least when compared to the much easier-to-grasp AE ad RAE. A second problem is that their typical values are usually difficult to make sense of, since genuinely engineered quantifiers may easily obtain values in $[10^{-6}, 10^{-2}]$.

A third, related problem is that realistic quantifiers trained by genuinely engineered quantification methods may obtain values that are different by orders of magnitude, which is something that experimenters may find difficult to interpret. As an example, assume a (very realistic) scenario in which $|\sigma| = 1000$, $\mathcal{Y} = \{y_1, y_2\}$, $p(y_1) = 0.01$, and in which three different quantifiers $\hat{p}'$, $\hat{p}''$, $\hat{p}'''$ are such that $\hat{p}'(y_1) = 0.0101$, $\hat{p}''(y_1) = 0.0110$, $\hat{p}'''(y_1) = 0.0200$. In this scenario KLD ranges on $[0, 7.46]$, $\text{KLD}(p, \hat{p}') = 4.78\text{e-}07$, $\text{KLD}(p, \hat{p}'') = 4.53\text{e-}05$, $\text{KLD}(p, \hat{p}''') = 3.02\text{e-}03$, i.e., the difference between $\text{KLD}(p, \hat{p}')$ and $\text{KLD}(p, \hat{p}'')$ and the difference between $\text{KLD}(p, \hat{p}'')$ and $\text{KLD}(p, \hat{p}''')$ are 2 orders of magnitude each, while the difference between $\text{KLD}(p, \hat{p}')$ and $\text{KLD}(p, \hat{p}''')$ is no less than 4 orders of magnitude. The increase in error (as computed by KLD) deriving from using $\hat{p}'''$ instead of $\hat{p}'$ is +632599%. We should add that, if (as noted at the beginning of Section 3.1) one wanted to average KLD results across a set of samples, the average would be completely dominated by the value with the highest order of magnitude, and the others would have little or no impact.

Unfortunately, switching from KLD to NKLD does not help much in this respect since, for realistic quantification systems, $\text{NKLD}(p, \hat{p}) \approx \frac{1}{2} \text{KLD}(p, \hat{p})$. The reason is that NKLD is obtained by applying a sigmoidal function (namely, the logistic function) to KLD, and the tangent to this sigmoid for $x = 0$ is $y = \frac{1}{2}x$; since the values of KLD for realistic quantifiers are (as we have observed above) very close to 0, for these values the $\text{NKLD}(p, \hat{p})$ curve is well approximated by $y = \frac{1}{2} \text{KLD}(p, \hat{p})$. As a measure for evaluating SLQ, NKLD thus *de facto* inherits most of the problems of KLD.

3.1.6   Which Measure is the Best for SLQ?

Figure 3.1 (adapted from Sebastiani (2020)) plots the six main evaluation measures discussed in Sections 3.1.3 to 3.1.5 for the binary case. Table 3.2 summarises instead, in compact form, the properties that these measures enjoy. From this table it appears evident that no measure proposed so far is completely satisfactory. Which measure should one adopt then?

KLD and NKLD are the least satisfactory ones, and seem out of the question. Concerning the others, the problem is that **MAX** seems to be incompatible with

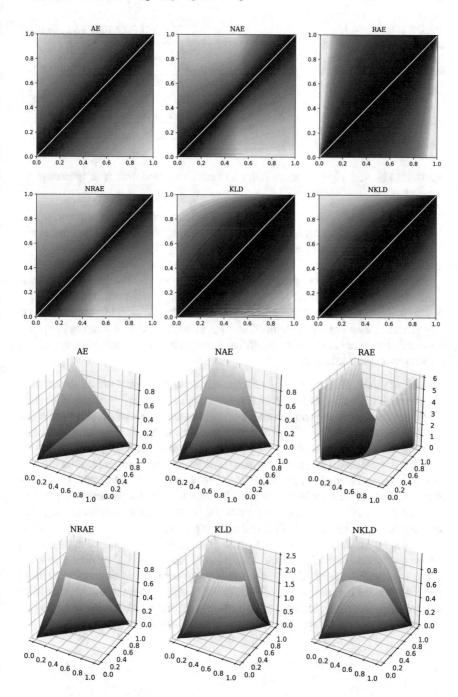

Fig. 3.1 2D plots and 3D plots (for a binary quantification task) for the six main evaluation measures mentioned in Sections 3.1.3 to 3.1.5; $p(y_1)$ and $p(y_2)$ are represented as x and $(1-x)$, respectively, while $\hat{p}(y_1)$ and $\hat{p}(y_2)$ are represented as y and $(1-y)$. Darker areas represent values closer to 0 (i.e., smaller error) while lighter areas represent values more distant from 0 (i.e., higher error).

Table 3.2 Properties of the evaluation measures for quantification discussed in this book. (Adapted from Sebastiani (2020)).

	MAX	IMP	REL	ABS
AE	No	✓	No	✓
NAE	✓	✓	No	No
RAE	No	✓	✓	No
NRAE	✓	✓	No	No
KLD	No	No	No	No
NKLD	✓	No	No	No

REL / ABS, and vice versa. In order to break the deadlock, it is important to remember that

1. the argument in favour of **REL** is that it reflects the needs of applications in which an estimation error of a given absolute magnitude should be considered more serious if it affects a rarer class;
2. the argument in favour of **ABS** is that it reflects the needs of applications in which an estimation error of a given absolute magnitude should be considered to have the same impact independently of the true prevalence of the affected class;
3. the main (although not the only) argument in favour of **MAX** is that, if an evaluation measure for quantification does not satisfy it, the n samples on which we may want to compare our quantification algorithms will each have a different weight on the final result.

Sebastiani (2020, §5.1) contends that Arguments 1 and 2 seem more important than Argument 3, since they are really about how an evaluation measure reflects the needs of the application; if the corresponding properties are not satisfied, one may argue that the quantification accuracy (or error) being measured is only loosely related to what the user really wants. Argument 3, while important, only implies that, if **MAX** is not satisfied, (1) results obtained on codeframes of different cardinality will not be comparable, and (2) results obtained on samples characterised by different true distributions will not be comparable. Despite this, results obtained by different systems on the same set of samples, even if this set contains samples that refer to codeframes of different cardinality, remain comparable.

This suggests that AE and RAE (or their "squared" versions, such as the SE measure of Section 3.1.3) are the measures of choice; AE should be preferred when an estimation error of a given absolute magnitude should be considered more serious when the true prevalence of the affected class is lower, while RAE should be chosen when an estimation error of a given absolute magnitude has the same impact independently of the true prevalence of the affected class.

3.2 Measures for Evaluating OQ

3.2.1 Earth Mover's Distance

The most popular measure for evaluating *ordinal quantification* systems is currently the *Earth Mover's Distance* (EMD – Rubner et al., 1998). EMD, also known as the *Vaseršteĭn metric* (Rüschendorf, 2001), is a function often used in content-based image retrieval for computing the distance between colour distributions of two images (see Levina and Bickel, 2001 for a rigorous probabilistic interpretation of the EMD). It was first proposed as an evaluation measure for ordinal quantification in Esuli and Sebastiani (2010b), and was used as the official evaluation measure of Subtask E "Tweet Quantification on a 5-point Scale" of SemEval-2016 and SemEval-2017 "Task 4: Sentiment Analysis in Twitter" (Nakov et al., 2016, 2017).

To see the intuition upon which the EMD is based, if the two distributions are interpreted as two different ways of scattering a certain amount of "earth" across different "heaps", the EMD is defined to be the minimum amount of work needed for transforming one set of heaps into the other, where the work is assumed to correspond to the sum of the amounts of earth moved times the distance by which they are moved. EMD may be seen as computing the minimal "cost" incurred in transforming one distribution into the other, where the cost is computed as the probability mass that needs to be shuffled around from one class to another, weighted by the "distance" between the classes involved.

Originally, the EMD is defined for the general case in which a distance $d(y', y'')$ is defined on $\mathcal{Y}^2$. In the much more specific case in which (a) there is a total order $y_1 \prec \ldots \prec y_{|\mathcal{Y}|}$ on the classes in $\mathcal{Y}$, and (b) $d(y_i, y_j) = d(y_i, y_{i+1}) + d(y_{i+1}, y_{i+2}) + \ldots + d(y_{j-2}, y_{j-1}) + d(y_{j-1}, y_j)$ for all $1 \leq i < j \leq |\mathcal{Y}|$ (as is the case in ordinal quantification), EMD comes down to the *Normalised Match Distance* (NMD) (Sakai, 2018; Werman et al., 1985), defined as

$$\text{NMD}(p, \hat{p}) = \frac{1}{|\mathcal{Y}| - 1} \sum_{j=1}^{|\mathcal{Y}|-1} d(y_j, y_{(j+1)}) \cdot | \sum_{i=1}^{j} \hat{p}(y_i) - \sum_{i=1}^{j} p(y_i)| \qquad (3.12)$$

where $\frac{1}{|\mathcal{Y}|-1}$ is just a normalisation factor that allows NMD to range between 0 (best) and 1 (worst).

The rationale of Equation 3.12 is the following. Assume that, in line with the interpretation of the NMD we have given above, in order to transform the estimated distribution $\hat{p}$ into the true distribution p we need to move some estimated probability across classes, from the ones where prevalence has been overestimated to the ones where prevalence has been underestimated. We formalise this by saying that if $p(y_i)$ has been overestimated there is going to be a positive quantity $(\hat{p}(y_i) - p(y_i))$ outgoing from class y_i, while if $p(y_i)$ has been underestimated this outgoing quantity is negative (which means that the incoming quantity is positive). In order to minimise the travelled distance, it makes sense to transfer probability mass from classes that are next to each other in the total order. The first step, from left to right, is thus to transfer $|\hat{p}(y_1) - p(y_1)|$ from y_1 to y_2 if $(\hat{p}(y_1) - p(y_1))$ is positive, or to transfer $|\hat{p}(y_1) - p(y_1)|$ from y_2 to y_1 if it is negative; in either case the cost of this transfer is $d(y_1, y_2) \cdot |\hat{p}(y_1) - p(y_1)|$. Since $\hat{p}(y_1)$ has now been transformed into $p(y_1)$, the next step is to transfer probability mass from y_2 to y_3. The probability mass outgoing from y_2 is now $(\hat{p}(y_1) + \hat{p}(y_2)) - (p(y_1) + p(y_2))$, which is going to be positive if y_1 and y_2 have altogether been overestimated, and negative otherwise; in either case the cost of this transfer is $d(y_2, y_3) \cdot |(\hat{p}(y_1) + \hat{p}(y_2)) - (p(y_1) + p(y_2))|$. Proceeding in this fashion, $|\mathcal{Y}| - 1$ probability mass transfers are performed, which explains Equation 3.12; this also shows that, in the form of Equation 3.12, NMD can be computed in $|\mathcal{Y}| - 1$ steps from the estimated and true class prevalence values.

Note that in many practical cases it happens that $d(y_i, y_{i+1}) = 1$ for all $i \in \{1, \ldots, (|\mathcal{Y}| - 1)\}$, which means that Equation 3.12 simplifies even further.

NMD can be seen as the ordinal equivalent of absolute error; in fact, assuming that $d(y', y'')$ is the same for all $y', y'' \in \mathcal{Y}^2$ (in which case ordinal quantification defaults to standard single-label quantification), the probability mass that needs to be moved from one class to another, weighted by the distance between the two classes (though this weighting is inessential, as all inter-class distances are the same) in order to recover p from $\hat{p}$, is exactly absolute error.

3.2.2 Root Normalised Order-Aware Divergence

Another proposed measure for evaluating the quality of OQ estimates is the *Root Normalised Order-aware Divergence* (RNOD), proposed by Sakai (2018) and defined as

$$\text{RNOD}(p, \hat{p}) = \left(\frac{\sum_{y_i \in \mathcal{Y}^*} \sum_{y_j \in \mathcal{Y}} d(y_j, y_i)(p(y_j) - \hat{p}(y_j))^2}{|\mathcal{Y}^*|(|\mathcal{Y}| - 1)} \right)^{\frac{1}{2}} \tag{3.13}$$

where $\mathcal{Y}^* = \{y_i \in \mathcal{Y} | p(y_i) > 0\}$.

However, RNOD is a more controversial measure for OQ than NMD since, without making it explicit, it penalizes more heavily mistakes (i.e., "transfers" of probability mass from a class to another) closer to the extremes of the codeframe. For instance, given codeframe $\mathcal{Y} = \{y_1, y_3, y_3, y_4, y_5\}$, assume that the true distribution is $p = (0.2, 0.2, 0.2, 0.2, 0.2)$, and assume two predicted distributions $\hat{p}' = (0.2, 0.2, 0.3, 0.1, 0.2)$ and $\hat{p}'' = (0.2, 0.2, 0.2, 0.3, 0.1)$. The two predicted distributions make essentially the same mistake, i.e., erroneously "transfer" a probability mass of 0.1 from a class y_i to a class $y_{(i-1)}$, the difference being that in $\hat{p}'$ it is the case that $i = 4$ and in $\hat{p}''$ it is the case that $i = 5$. NMD penalizes them equally (since $\text{NMD}(p, \hat{p}') = \text{NMD}(p, \hat{p}'') = 0.1$). RNOD instead does not (since $\text{RNOD}(p, \hat{p}') \approx 0.080$ while $\text{RNOD}(p, \hat{p}'') \approx 0.092$), and the degree to which mistakes closer to the extremes of the codeframe are penalised more heavily than the ones close to the center of the codeframe, is not explicit in the formula.

Other OQ evaluation measures are proposed by Sakai (2021), such as *Root Symmetric Normalised Order-aware Divergence* (RSNOD) and *Root Normalised Average Distance-Weighted sum of squares* (RNADW), but we do not consider them here since they are variants of RNOD that share the characteristics of RNOD mentioned above.

3.3 Measures for Evaluating Regression Quantification

The only work to date that investigates *regression quantification* (RQ) is Bella et al. (2014); this work (discussed in details in Section 5.2) is thus also the only one that discusses how to perform evaluation for this task.

In a regression problem, every input object is assigned with a real-valued score as output, differently from the classification case in which the output has a categorical form. For the regression quantification scenario, Bella et al. (2014) identify two possible quantification goals, aggregated indicator estimation and distribution estimation.

The case of estimating an aggregated indicator is defined as the one in which the interest is on estimating a statistic function $\mathbb{I}$ that summarises some property of the distribution on regression scores over the unlabelled set U. A typical example of indicator function in regression quantification is the average of the regression values of the elements, i.e.,

$$\mathbb{I}(U) = \mu_U$$

$$= \frac{1}{|U|} \sum_{i=1}^{|U|} y_i \tag{3.14}$$

Bella et al. (2014) observe that the single numerical value produced by an aggregated indicator can be compared to the true value from test data using an error measure such as the Squared Error (see Section 3.1.3).

The optimal value of SE is always zero, while the upper bound of this measure depends on the range of regression values and distribution of data. Bella et al. (2014) propose a variant of SE, VSE, that normalises the SE value by the variance of the training set, i.e.,

$$\text{VSE}(p, \hat{p}, L) = \frac{\text{SE}(p, \hat{p})}{\text{Var}(L)} \tag{3.15}$$

where $\text{Var}(L)$ is the variance of the true regression scores computed on elements in the training set. The motivation the authors give to support VSE is to make the results from experiments less dependent on the magnitude range of the data when such experiments are run on different datasets, or involve repeated runs.

Bella et al. (2014) do not provide a theoretical motivation to support VSE, in particular on why the variance should be computed on the training set, instead of the test set. In experiments using different test sets, either from different datasets or produced by sampling, the variance of the test set has the same ability to measure the difference in the magnitude range of values. In experiments using the same test set, a training set with higher variance get an advantage when evaluated using VSE. The intuition the authors followed may be that learning from a training set with higher variance is more difficult than learning from a low variance one. Yet, this does not take into account the actual variance of the test set. It could be the case that also the test set has a high variance, and the lower variance training set is thus the one from which is more difficult to learn an accurate regressor. Moreover, it is possible to cheat VSE by adding dummy examples in training set with the sole aim of increasing variance.

The case of estimating a probability distribution over the regression values can be evaluated comparing distribution of the true values with the one of predicted values. This is an intermediate step in the complexity of prediction between accurately predicting each single value, i.e., the actual regression problem, and predicting an aggregated estimator.

A way to compare probability distributions is to use divergence measures, yet Bella et al. (2014) observe that some divergence measures are not always defined when comparing empirical distributions[5]. and they thus suggest to perform the evaluation using the cumulative distributions. Within the set of measures that compare cumulative distributions they mention the Kolmogorov-Smirnov statistic measures, but they criticise the fact that it only considers the point where the distributions differ the most, when the entire shape of the distributions should be considered instead. For this reason they thus suggest to adopt, as a better refined evaluation measure, the *Cramér-von-Mises statistic* (Anderson, 1962) that computes an integral between the difference of the two cumulative distributions. More specifically, they adopt in their experiments the *L1*-version of the statistic (Xiao et al., 2006).

3.4 Experimental Protocols for Evaluating Quantification

Any test set routinely used for testing the accuracy of classification can obviously be used also for evaluating quantification. However, the problem is that, while for classification a set of k unlabelled data provides k unlabelled data points,

---

[5] E.g., Kullback-Leibler divergence requires that $p(x) = 0 \implies \hat{p}(x) = 0$. KL is thus undefined when the regressor predicts even a single value that is not within the set of values appearing in the test set.

for quantification the same test set just provides 1 test data point. Evaluating quantification algorithms is thus challenging, due to the fact that the availability of labelled data for testing purposes is not unlimited.

There are two main experimental protocols that have been taken in order to deal with this problem; we will here call them the *Natural-Prevalence Protocol* (NPP) and the *Artificial-Prevalence Protocol* (APP).

3.4.1 Natural Prevalence Protocol (NPP)

The NPP was first used by Esuli and Sebastiani (2015). It consists of taking a large enough test set, partitioning it in a number of samples, and carrying out the evaluation individually on each such sample. For instance, Esuli and Sebastiani (2015) tested binary quantifiers on the well-known RCV1-v2 text classification dataset, whose test set consists of about 780,000 news items issued by the Reuters news agency over a period of 52 weeks, and labelled with 99 different classes. This allowed the authors to split the test set in 52 samples (each corresponding to a week), each of which provided 1 testing data point for 99 different BQ experiments, thus generating $52 \times 99 = 5148$ testing data points.

3.4.2 Artificial Prevalence Protocol (APP)

The APP was first used by Saerens et al. (2002). This protocol consists of taking a standard dataset, split into a training set L and a set U of unlabelled items, and conducting repeated experiments in which either the training set prevalence or the test set prevalence of a class are artificially varied via subsampling. For instance, in the BQ experiments carried out by Forman (2005), given codeframe $\mathcal{Y} = \{\oplus, \ominus\}$, repeated experiments are conducted in which either examples of $\oplus$ or examples of $\ominus$ are removed at random from the test set in order to generate a predetermined prevalence of $\oplus$ in the sample U thus obtained. In this way, different samples can be generated, each characterised by a different prevalence of $\oplus$ (e.g., $p_U(\oplus) \in \{0.00, 0.05, \ldots, 0.95, 1.00\}$). This can be repeated, thus generating multiple random samples for each class prevalence. Analogously, random removal of either positive or negative examples can be performed on the training set, thus bringing about training sets with different values of $p(\oplus)$. Example results of the application of the APP will be illustrated in Section 6.3.

Doing an analogous grid-based exploration in the SLQ setting is certainly possible, but cumbersome; for instance, if we want to restrict ourselves to class prevalence values in the set $\{0.00, 0.05, \ldots, 0.95, 1.00\}$, there are just 21 possible distributions in the BQ case, but in the SLQ case there are many, many more, especially when $|\mathcal{Y}|$ is high, due to combinatorial explosion. If we use a grid of class prevalence values $g = \{\frac{i}{m}\}_{i=0}^{m}$ containing $|g| = m + 1$ possible values (with m an integer), and we have $|\mathcal{Y}| = 2$ classes, then there are $(m + 1)$ choices for $p_\sigma(y_1)$; of course, these constrain the value of $p_\sigma(y_2)$, which must be equal to $p_\sigma(y_2) = 1 - p_\sigma(y_1)$.

Let us define a function $K(m, n)$ that computes the number of possible combinations for $|\mathcal{Y}| = n$ classes using a grid of prevalence values, from 0 to 1 at a step size of $\frac{1}{m}$. For the binary case discussed above, it is the case that $K(m, 2) = m + 1$. For the ternary case, i.e., when $n = 3$, we have $K(m, 3) = (m + 1)(m + 2)/2$. This follows from the observation that, when we set $p_\sigma(y_1) = 0/m$, there are $m + 1$ possible choices for $p_\sigma(y_2)$ (while $p_\sigma(y_3) = 1 - (p_\sigma(y_1) + p_\sigma(y_2))$ is constrained); when we set $p_\sigma(y_1) = 1/m$, there only exist m possible choices for $p_\sigma(y_2)$; and so on, until we end up setting $p_\sigma(y_1) = m/m = 1$, for which there is only one possible combination $p_\sigma(y_2) = p_\sigma(y_3) = 0$ representing a valid distribution. In our previous example, with $|g| = 21$, we thus have $K(20, 3) = 231$. In general, for arbitrary m and n values, the number of possible prevalence distributions can be derived from the so-called "stars and bars" method[6] and is given by

$$K(m, n) = \binom{m + n - 1}{n - 1} \tag{3.16}$$

3.4.3 A Variant of the APP Based on the Kraemer Algorithm

As one would expect by looking at Equation 3.16, the number of possible distribution vectors that the APP generates grows very rapidly. To exemplify, for 5 classes we already reach $K(20, 5) = 10{,}626$ valid combinations, while for 10 classes the number of combinations rises to $K(20, 10) = 10{,}015{,}005$. Things get even worse when using a finer-grained grid; for example, using a stepsize of 0.01 (i.e., setting $m = 100$) the number of combinations to explore for 10 classes, $K(100, 10) > 4E^{12}$, becomes impractical.

One possible solution consists of simply renouncing to *predetermine* class prevalence values, and instead letting them vary at random, by first generating a random distribution p and then generating a sample σ by randomly picking items

---

[6] A probability distribution of n classes taking prevalence values from a grid g of $(m+1)$ prevalence values of probability mass $1/m$ each, can be seen as a vector of $(m + n - 1)$ positions filled with m "stars" (each star representing a probability mass of $1/m$) and $(n - 1)$ "bars" (each bar representing a separator for two adjacent classes). For example, for $n = 4$ and $m = 10$, the probability distribution given by $p_\sigma(y_0) = 0.3$, $p_\sigma(y_1) = 0$, $p_\sigma(y_2) = 0.6$, and $p_\sigma(y_3) = 0.1$, corresponds to the vector of "stars and bars" $(*, *, *, |, |, *, *, *, *, *, *, |, *)$, where each '$*$' amounts to 0.1 of probability mass, and there are $(n - 1)$ separators '$|$'. The number of ways $(n - 1)$ bars (resp., m stars) can be inserted in a vector of $(m + n - 1)$ positions, with the remaining elements set to stars (resp., bars) is given by the binomial coefficient above. See https://brilliant.org/wiki/integer-equations-star-and-bars/#stars-and-bars for further details.

from the population according to p. Somehow unexpectedly though, sampling distribution vectors p *uniformly* at random, i.e., so that all legitimate distribution vectors are equally likely, is not a trivial task. An intuitive and straightforward procedure, consisting of drawing n values, uniformly at random from the $[0, 1]$ interval, and then normalizing each value by the sum of all values (the sampling method used in Esuli et al. (2021)), corresponds to a sampling distribution that is strongly biased towards the centre of the distribution, for reasons that are discussed by Smith and Tromble (2004). Luckily enough, Smith and Tromble (2004) presented also a correct sampling algorithm, called the *Kraemer algorithm*, for sampling the unit $(n - 1)$-simplex[7] uniformly. Given a set n of classes, the method works as follows:

1. Generate a vector $\mathbf{x} = (x_1, \ldots, x_{(n-1)})$ of values, where each x_i is sampled uniformly at random from $[0,1]$
2. Sort $\mathbf{x}$ to obtain $\mathbf{x}'$, so that $x_1' \leq \ldots \leq x_{(n-1)}'$, and define two additional values $x_0' = 0$ and $x_n' = 1$
3. Return the distribution vector $p = (p_1, \ldots, p_n)$ in which $p_i = (x_i' - x_{(i-1)}')$ for every $i \in \{1, \ldots, n\}$

The Kraemer sampling algorithm has two additional advantages with respect to sampling based on a predefined grid: (i) it allows the practitioner to draw a desired number of samples, instead of imposing to generate all $K(m, n)$ valid combinations from the grid of prevalence values; and (ii) it truly allows any possible distribution vector to be picked, while this is not possible when using a grid of values, and especially so when the grid is a coarse-grained one.

To the best of our knowledge, the first experimental setting in the quantification literature that adopts the Kraemer algorithm as the sample-generating function is the one described in Esuli et al. (2022). Since this work is very recent, the version of the APP that is generally used by the quantification community is the "grid-based" version that we have discussed above. It remains to be see whether the version that adopts the Kraemer algorithm will gain significant acceptance in the years to come.

3.4.4 Should we Use the NPP or the APP?

The APP is much more widely used than the NPP in the quantification literature, possibly due to the difficulty of finding the large enough test sets that the NPP requires. However, both protocols have different pros and cons. One advantage of the APP (and a corresponding disadvantage of the NPP) is that it allows many test data points to be created from the same test set; it is not always the case that test sets large enough for the NPP to be adopted (such as the above-mentioned RCV1-v2)

---

[7] A distribution vector $p = (p_\sigma(y_1), \ldots, p_\sigma(y_n))$ belongs to the unit $(n - 1)$-simplex since $p_\sigma(y_i) \in [0, 1]$ for all $y_i \in \mathcal{Y}$ and since $\sum_{y_i \in \mathcal{Y}} p_\sigma(y_i) = 1$.

are available, so, when a smaller test set is all we have, the APP allows generating test data points almost at will. Additionally, the APP allows many situations (i.e., different training class prevalence values, different test class prevalence values, different amounts of shift, ...) to be simulated; in such a way, one can test the robustness of a quantification algorithm on many conditions even if the dataset itself does not naturally exhibit such conditions. However, one disadvantage of the APP is that it is not clear how realistic these different situations are; e.g., if $p(y)$ in the test set is 0.05, testing a quantifier on a sample U extracted from it such that $p_U(y) = 0.95$ might be challenging but unrealistic, since these amounts of shift may be unlikely in real-world applications. The NPP, by focusing on real samples and really occurring situations, scores higher than the APP in terms of realism.

It may be worth noting that some of the problems discussed above might be solved by defining a protocol "intermediate" between the NPP and the APP, i.e., a protocol which uses prior knowledge about the distribution of "likely" prevalence vectors that one could expect to encounter in the specific domain at hand. However, we are unaware of previous experiments that used this or similar approaches, likely due to the fact that, in real scenarios, it is difficult to have any such prior knowledge about how the distribution might vary. Anyway, the bottom line is that the pros and cons of the APP and the NPP have led to some controversy around the adequacy of these protocols in the assessment of the performance of quantification systems. Hassan et al. (2021) expose the shortcomings and potential risks that the adoption of the APP (more specifically: of the grid-based variant discussed in Section 3.4.2) might bring about in the evaluation. Among other things, the authors reported that knowing in advance the expected value of the distribution vectors that the APP generates (e.g., that in binary quantification the positive class has an expected prevalence value of $\mathbb{E}[p_\sigma(\oplus)] = 0.5$) might be maliciously exploited in order to get an (illusory) advantage over other methods that do not make such assumption.

It should also be mentioned that the APP, as it has been used up to now, only models either covariate shift or prior probability shift, and does not model concept shift. To see this, assume we are dealing (see Section 1.5) with a "$\mathcal{X} \rightarrow \mathcal{Y}$" problem, which is modelled by Equation 1.3. By subsampling the test set we are simulating covariate shift, whereby $p(y)$ changes only because $p(\mathbf{x})$ changes (since we have selectively removed specific data items $\mathbf{x}$); note that $p(y|\mathbf{x})$ does not change, i.e., there is no concept shift, since the labels of the items that have not been removed have remained the same. Concept shift could be simulated not by removing labelled items but by flipping the labels on some data items, e.g., according to one of the two methods discussed in Esuli and Sebastiani (2013). With this method, given a set U of unlabelled items, many samples that contain the very same data items (which means there is no covariate shift) could be generated by flipping different subsets of the items contained in U. Conversely, assume we are dealing with a "$\mathcal{Y} \rightarrow \mathcal{X}$" problem, which is modelled by Equation 1.4. By subsampling the test set we are simulating prior probability shift, whereby $p(y)$ changes *motu proprio* (since we have selectively removed data items characterised by a specific label y); note that $p(\mathbf{x}|y)$ does not change, i.e., there is no shift in the within-class densities, since the feature vectors of the items that have not been removed have remained the same.

3.5 Model Selection in Quantification

The performance of many machine learning algorithms depends on how their *hyperparameters* are set. Hyperparameters control specific aspects of the learning process and, in contrast to the *parameters* of the model, they are not learned during the training phase, but are instead set in advance.

Although machine learning methods often come with default values for the hyperparameters (values that the inventors of the method have found to work reasonably well in a variety of scenarios), it is well known that the final performance can often be improved by carefully tuning the hyperparameters for the specific applicative domain. Quantification systems are no exception in this regard.

The process of hyperparameter optimisation is known as *model selection*, and typically consists of testing how well the model fares when setting the hyperparameters with different combinations of values from a set of candidate configurations. Model selection is carried out in a fully automated way, i.e., the model's performance is assessed on held-out validation data or via cross-validation.

Model selection is thus inherently related to performance evaluation. Hyperparameter optimisation should thus mimic the evaluation protocol (using validation data) when assessing the adequacy of each of the candidate configurations. Since quantification has become a task on its own right, with dedicated evaluation measures (Sections 3.1, 3.2, and 3.3) and dedicated experimental protocols (Section 3.4), it should likewise have specific model selection routines (Moreo and Sebastiani, 2021). In other words, since the goal of model selection is to choose the configuration of hyperparameters that perform best according to a given experimental protocol and a given evaluation measure, it makes perfect sense to adopt the same evaluation protocols and error metrics customarily used in the evaluation of quantification systems.

Somehow surprisingly, though, the quantification community has largely overlooked this aspect in the past. In a large body of quantification work, it is not even documented whether the hyperparameters were optimised at all (Esuli and Sebastiani, 2014; Forman, 2008; González et al., 2017; González-Castro et al., 2013; Hopkins and King, 2010; Levin and Roitman, 2017; Pérez-Gállego et al., 2017; Saerens et al., 2002). Other papers Barranquero et al. (2015); Bella et al. (2010); Esuli and Sebastiani (2015); Hassan et al. (2020); Milli et al. (2013) simply report that the hyperparameters were left to their default values; others do not document the evaluation measure being optimised during model selection (Esuli et al., 2020; Gao and Sebastiani, 2016), or instead optimise for a classification-oriented loss (Barranquero et al., 2013; Pérez-Gállego et al., 2019).

The only paper we are aware of that proposed the use of a quantification-oriented optimisation of hyperparameters is Moreo and Sebastiani (2021). In their work, the authors claimed that the "Classify and Count" (CC – Section 4.2.1) method and its variants (Sections 4.2.2, 4.2.3, 4.2.4), routinely used as baseline models in experimental evaluations, have largely been misrepresented, since they have never been optimised properly for the task of quantification. In their results they showed that, when properly optimised, these simple method become respectable contenders, even if still inferior to the most sophisticated quantification methods.

Chapter 4
Methods for Learning to Quantify

This section is devoted to discussing methods that have been proposed in the literature for tackling quantification[1]. All of these methods rely on supervised learning, and depart from standard classification methods in one or more ways.

As in the rest of this book, our main focus is (for the reasons discussed in Section 1.4) single-label multiclass quantification. While many of the methods that will be discussed in this section can natively deal with the single-label multiclass case, some other methods (for example, those of Sections 4.2.5, 4.2.12 and 4.3.1) are only defined for the binary case, and cannot easily be extended to the single-label multiclass case. In order to use them for single-label multiclass quantification, it is thus necessary to run them in binary mode for each class in the codeframe, and to normalise the resulting class prevalence values so that they sum to 1.

Broadly speaking, two large classes of methods can be discerned in the literature.

The first class is that of *aggregative methods*, i.e., methods that require the classification of all the individual data items as an intermediate step; these methods will be the subject of Sections 4.2 and 4.3. Within the class of aggregative methods, two subclasses can be identified. The first subclass (Section 4.2) includes methods based on general-purpose learners; in these methods the classification of the individual items performed as an intermediate step may be accomplished by means of any classifier. The second subclass (Section 4.3) is instead composed of methods that, in order to classify the individual data items, rely on special-purpose learning methods devised with quantification in mind.

The second class (Section 4.3) is that of *non-aggregative* methods, i.e., methods that solve the quantification task "holistically", i.e., without classifying the individual items; these methods will be the subject of Section 4.4.

---

[1] Section 6.2 presents a lists of software tools implementing quantification methods, including many of those presented in this section.

© The Author(s) 2023
A. Esuli et al., *Learning to Quantify*, The Information Retrieval Series 47,
https://doi.org/10.1007/978-3-031-20467-8_4

We start by discussing, in the next section, a method that belongs to none of the classes above, but that is sometimes considered as a trivial baseline in comparative experiments.

4.1 Maximum Likelihood Prevalence Estimation

Maximum Likelihood Prevalence Estimation (MLPE) is not a real quantification method, but is sometimes used (see e.g., Barranquero et al., 2013) as a trivial baseline against which genuine quantification methods are compared. MLPE makes the naïve assumption that there is zero distribution shift between L and U, and thus consists of taking $p_L(y)$ as an estimate of $p_U(y)$, i.e.,

$$\hat{p}_U^{\mathrm{MLPE}}(y) = p_L(y) \tag{4.1}$$

This is the trivial predictor for quantification, somehow akin to always picking the majority class in classification.[2]

However, that MLPE should be used as a baseline at all is questionable. In fact, on a dataset where it indeed happens that $p_U(y) = p_L(y)$, MLPE cannot be beaten by any genuine quantification method, and will be hardly beaten by any such method on datasets characterised by very low distribution shift. However, this should not be taken to mean that these genuine quantification methods are ineffective, but should rather indicate that we have chosen the wrong dataset(s), since there is no point in applying quantification in environments characterised by the absence of distribution shift.[3]

As a side note, the assumption that $p_U(y) = p_L(y)$ has been used in the past to justify classification policies. For instance, in the binary case, Yang (2001) defines a strategy (called "Pcut") for optimising classification thresholds, which consists

---

[2] Given a prediction task, an effectiveness measure M for it, and labelled and sets of unlabelled items L and U (assumed to be independently and identically distributed), the *trivial predictor* may be defined as the predictor we obtain if we attempt to maximise M on U by using only the output variables (and not the input variables) of L. When "vanilla" accuracy (i.e., the fraction of classification decisions that are correct) is the effectiveness measure, the classifier that always predicts the majority class is the trivial predictor for both binary and multiclass classification; under any reasonable effectiveness measure, MLPE is the trivial predictor for quantification.

[3] As an example, assume we are asked by a customer to set up a system that monitors, in a stream of data, the class prevalence of a certain class y of interest to the customer. (For instance, the data may be textual comments about a product marketed by the customer, and $y = \oplus$ may be the class of positive such comments.) Assume also that the customer provides us with a training set L of comments labelled according to $\mathcal{Y} = \{\oplus, \ominus\}$, where $p_L(\oplus) = k$. If the system we deliver to the customer is one that always returns $p_\sigma(\oplus) = k$, for any sample σ that we may sample from the stream, the customer would not be happy, even if we justify this by saying that, on our test data, this system has outperformed any other genuine quantification system we have tested.

of picking the threshold that causes $p_U(y)$ to be equal (or as close as possible) to $p_L(y)$.

4.2   Aggregative Methods Based on General-Purpose Learners

In this section and in Section 4.3 we will discuss quantification methods that have an *aggregative* nature, i.e., that first require a (hard or soft) classifier to issue a prediction for each individual item, and that then output an estimated class prevalence based on these individual predictions. Indeed quantification- and classification-related goals can be supported, to some degree, by the same training strategy (Tasche, 2021). All the methods discussed in this section can be applied on top of any supervised learning algorithm for training classifiers.

Of the methods discussed in this section,

1. some (e.g., the ones of Sections 4.2.1 and 4.2.3) require as input the class labels (as returned by a hard classifier), while
2. some others (e.g., the ones of Sections 4.2.2 and 4.2.4) require as input the posterior probabilities (as output by a soft classifier).

For methods of type 2, these posterior probabilities should be well calibrated (in the sense discussed in Section 2.1). Some classifiers are known to return well calibrated probabilities (e.g., classifiers trained via logistic regression (Zadrozny and Elkan, 2002)). The posterior probabilities returned by some other classifiers are known instead to be not well calibrated (e.g., this is the case of the naïve Bayesian classifier (Domingos and Pazzani, 1997)). Yet some other classifiers (e.g., those trained via SVMs or AdaBoost) do not return posterior probabilities, but generic confidence scores. In these two last cases it is possible to map the obtained posterior probabilities / confidence scores into well calibrated posterior probabilities via some calibration method (Platt, 2000; Zadrozny and Elkan, 2002).

All this basically means that any supervised learning method can be used both for methods of type 1 and for methods of type 2 above. We now discuss all of these methods in increasing order of sophistication.

4.2.1   Classify and Count

An obvious method for quantification consists of training a hard classifier h from L via a standard learning algorithm, classifying the items in sample U, and estimating $p_U(y)$ by simply counting the fraction of items in U that are predicted to belong to class y. This corresponds to computing

$$\hat{p}_U^{CC}(y) = p_U^h(\hat{y})$$
$$= \frac{|\{\mathbf{x} \in U | h(\mathbf{x}) = y\}|}{|U|} \qquad (4.2)$$

Forman (2008) calls this the *Classify and Count* (CC) method.

As already discussed in Section 1.2, CC is sub-optimal, because standard classifiers might be biased, i.e., generate severely unbalanced numbers of false positives and false negatives, and because they are usually tuned to minimise a measure of classification error, and not of quantification error. However, CC plays an important role in quantification research since it is always used as the trivial baseline which any reasonable quantification method must improve upon.

4.2.2 Probabilistic Classify and Count

Probabilistic Classify and Count (PCC) is a variant of CC which consists of using L for training a probabilistic classifier $s : \mathcal{X} \rightarrow [0, 1]^{|\mathcal{Y}|}$, generating a posterior probability $p(y|\mathbf{x})$ for each item $\mathbf{x} \in U$ and for each class $y \in \mathcal{Y}$, and computing $p_U(y)$ as the *expected* fraction of items predicted to belong to y. If by $E[x]$ we indicate the expected value of x, this corresponds to computing

$$\hat{p}_U^{PCC}(y) = E[p_U^h(\hat{y})]$$
$$= \frac{1}{|U|} \sum_{\mathbf{x} \in U} p(y|\mathbf{x}) \qquad (4.3)$$

As a quantification method, PCC was first used by Bella et al. (2010), where it is called "Probability Average"[4]. The rationale of PCC is that posterior probabilities contain richer information than classification decisions, which are usually obtained from posterior probabilities via Equation 1.2. When using the classification decisions, CC does not leverage the quantitative information encoded in the $p(y|\mathbf{x})$'s, which is discarded when using Equation 1.2, and this may be suboptimal.

As a quantification method, PCC was first evoked by Lewis (1995), who stated that "(…) if our goal is to count class members, and if we have estimates of the probability of class membership, we should use the estimates directly to estimate the number of class members, rather than use them to classify documents." PCC was later dismissed *a priori* (i.e., without even being tested) as unsuitable by

---

[4] Tang et al. (2010) also use a method called "Probabilistic Classify and Count", and they also show that it outperforms CC. Their method might indeed coincide with the method discussed in this section, but the authors do not explain what their method precisely consists of.

Forman (2005, 2008), on the grounds that, when the training distribution p_L and the unlabelled distribution p_U are different (as they should be assumed to be in any application of quantification), probabilities calibrated on L (L being the only available set where calibration may be carried out) cannot be, by definition, calibrated for U at the same time (see also Section 2.1). Forman's criticism is indeed well-taken, since the assumption underlying the very notion of probability calibration is the IID assumption, whose consequence (namely, that class prevalence values are invariant across the training and the set of unlabelled items) is at odds with the very notion of quantification.

4.2.3 Adjusted Classify and Count

Adjusted Classify and Count (ACC – also called "Adjusted Count" in Forman (2008) and the "Confusion Matrix Method" in Saerens et al. (2002)) requires training a hard classifier h from L via a standard learning algorithm, classifying the items in U, and then observing that, thanks to the law of total probability, it holds that

$$p_U^h(\hat{y}_j) = \sum_{y_i \in \mathcal{Y}} p_U^h(\hat{y}_j | y_i) \cdot p_U(y_i) \tag{4.4}$$

Here, $p_U^h(\hat{y}_j | y_i)$ represents the fraction of data items in U whose true class is y_i and that have been instead assigned to class y_j by classifier h. Once the classifier has been trained and applied to U, the quantity $p_U^h(\hat{y}_j)$, which represents the fraction of items in U that have been assigned y_j by classifier h, can be observed, and the quantity $p_U^h(\hat{y}_j | y_i)$ can be estimated from L via k-fold cross-validation (k-FCV)[5]; the quantity $p_U(y_i)$ is instead unknown, and is indeed the quantity we want to estimate. Since there are $|\mathcal{Y}|$ equations of the type described in Equation 4.4 (one for each $y_j \in \mathcal{Y}$), and since there are $|\mathcal{Y}|$ quantities of type $p_U(y_i)$ to estimate (one for each $y_i \in \mathcal{Y}$), we are in the presence of a system of $|\mathcal{Y}|$ linear equations in $|\mathcal{Y}|$ unknowns. This system can be solved via standard techniques, thus yielding the $\hat{p}_U^{ACC}(y_i)$ estimates.

In a nutshell, ACC is based on the idea of adjusting the results of CC by taking into account the propensity of the classifier to make misclassifications of a certain type. This is particularly evident in the binary case $\mathcal{Y} = \{\oplus, \ominus\}$, where Equation 4.4

---

[5] Barranquero et al. (2013); Forman (2005, 2008) actually use *stratified* k-fold cross-validation, i.e., the training set is split in such a way as to ensure that the class distribution is invariant across the different folds. Given that our goal is quantification (i.e., that we assume the presence of distribution shift), the rationale of using stratification seems dubious here, given that we do not have any guarantee that the distribution that stratification enforces in the various folds will be the same in the test set.

comes down to

$$
\begin{aligned}
p_U^h(\hat{\oplus}) &= p_U^h(\hat{\oplus}|\oplus) \cdot p_U(\oplus) + p_U^h(\hat{\oplus}|\ominus) \cdot p_U(\ominus) \\
&= \mathrm{TPR}_U \cdot p_U(\oplus) + \mathrm{FPR}_U \cdot p_U(\ominus) \qquad (4.5) \\
&= \mathrm{TPR}_U \cdot p_U(\oplus) + \mathrm{FPR}_U \cdot (1 - p_U(\oplus))
\end{aligned}
$$

where by $\mathrm{TPR} = \frac{\mathrm{TP}}{\mathrm{TP+FN}}$ and $\mathrm{FPR} = \frac{\mathrm{FP}}{\mathrm{FP+TN}}$ we indicate the true positive rate (a.k.a. "recall", or "sensitivity") and false positive rate (a.k.a. "specificity"), resp., that the classifier has obtained. From Equation 4.5 we obtain

$$
\begin{aligned}
p_U(\oplus) &= \frac{p_U^h(\hat{\oplus}) - \mathrm{FPR}_U}{\mathrm{TPR}_U - \mathrm{FPR}_U} \\
&= \frac{\hat{p}_U^{\mathrm{CC}}(\oplus) - \mathrm{FPR}_U}{\mathrm{TPR}_U - \mathrm{FPR}_U}
\end{aligned} \qquad (4.6)
$$

from which, if by TPR_L and FPR_L we indicate the true positive rate and false positive rate, resp., that have been estimated by k-FCV, we derive

$$
\hat{p}_U^{\mathrm{ACC}}(\oplus) = \frac{\hat{p}_U^{\mathrm{CC}}(\oplus) - \mathrm{FPR}_L}{\mathrm{TPR}_L - \mathrm{FPR}_L} \qquad (4.7)
$$

ACC can be proved to be *Fisher-consistent* under prior probability shift (Tasche, 2017), which is a guarantee that the provided estimate $\hat{p}_U(y)$ would be correct if computed on the whole populations of interest (instead of the available samples L and U of limited cardinality), on condition that the training and unlabelled populations are linked by *prior probability shift*.

Fisher consistency is related to an estimator being unbiased, and has been proposed as a desirable property of a quantification method (Tasche, 2017). Fisher consistency does not provide any practical guarantee, given it discounts the randomness of empirical distributions sampled in the real world. However, a quantification method lacking this property can be seen as problematic, given that, even for large sample size, it may end up providing poor estimates of class prevalence. Thus it can be seen as a necessary, not sufficient, property of a good quantification method. Dataset shifts (Section 1.5) close to, but slightly deviating from, prior probability shift can cause a loss of Fisher consistency for ACC (Tasche, 2017).

One problem with ACC is that the $\hat{p}_U^{\mathrm{ACC}}(y_i)$'s are not guaranteed to be in [0,1], due to the fact that the estimates of the $p_U^h(\hat{y}_j|y_i)$'s may be inaccurate, i.e., substantially different from the true $p_U^h(\hat{y}_j|y_i)$'s.[6] In fact, ACC is based on the hypothesis that the $p_U^h(\hat{y}_j|y_i)$'s are invariant across the training set and the set of

---

[6] This problem had already been noted by Lew and Levy (1989).

unlabelled items, which is questionable in the presence of distribution shift.[7] The fact that the $\hat{p}_U^{ACC}(y_i)$'s may not be in [0,1] particularly affects classes characterised by a low or very low prevalence (which are ubiquitous in e.g., text classification): in these case it may well be that $\hat{p}_U^{CC}(y) < \text{FPR}_L$, which means that, since in these scenarios it is usually the case that $\text{TPR}_L > \text{FPR}_L$, ACC returns a negative value.

This problem has led most authors (see e.g., Forman, 2008) to rely on "clipping and rescaling", i.e., (i) "clip" the $\hat{p}_U(y_i)$ estimates (i.e., equate to 1 every value higher than 1 and to 0 every value lower than 0), and (ii) rescale them so that they sum up to 1. Clipping is a hardly justified heuristics, though, and if the values to be clipped are either much smaller than 0 or much higher than 1, it can seriously bias the results. A better alternative (that does away with clipping, but that – to the best of our knowledge – has never been discussed in the literature) might consist of giving the $p_U^h(\hat{y}_j)$ values obtained from Equation 4.4 as input to a "softmax" function

$$\sigma(x) = \frac{e^x}{\sum_{x_i} e^{x_i}} \tag{4.8}$$

whose effect is to monotonically map the $\hat{p}_U(\hat{y}_j)$'s obtained from solving the system of linear equations, to $|\mathcal{Y}|$ values in [0,1] that sum up to 1. The values returned by the softmax would then be used as the final $\hat{p}_U^{ACC}(\hat{y}_j)$ class prevalence estimates in place of the values computed from the system of linear equation.

ACC is actually very old, since its binary version goes back at least to (Gart and Buck, 1966), where, in an application pertaining to epidemiology, it was used in order to determine the prevalence of a given disease from the results of a screening test with known true positive rate and true negative rate (see Section 6.4 for more on this).[8] As a quantification method, the earliest recorded use of it is in Vucetic and Obradovic (2001).

4.2.4 Probabilistic Adjusted Classify and Count

Probabilistic Adjusted Classify and Count (PACC) is a probabilistic variant of ACC, i.e., it stands to ACC as PCC stands to CC. Its underlying idea is to replace both side

---

[7] Esuli and Sebastiani (2015, Appendix A) show an example in which this assumption is far from being verified in actual data.

[8] Several past works erroneously attribute this method to Levy and Kass (1970); in reality, the latter paper does use the method, but the authors correctly attribute its paternity to Gart and Buck (1966).

of Equation 4.4 with their expected values. Equation 4.4 is thus transformed into

$$E[p_U^h(\hat{y}_j)] = E[\sum_{y_i \in \mathcal{Y}} p_U^h(\hat{y}_j|y_i) \cdot p_U(y_i)]$$

$$= \sum_{y_i \in \mathcal{Y}} E[p_U^h(\hat{y}_j|y_i) \cdot p_U(y_i)] \tag{4.9}$$

$$= \sum_{y_i \in \mathcal{Y}} E[p_U^h(\hat{y}_j|y_i)] \cdot p_U(y_i)$$

where the last passage is justified by the fact that $p_U(y_i)$ is a constant, and where

$$E[p_U^h(\hat{y}_j)] = \frac{1}{|U|} \sum_{\mathbf{x} \in U} p(y_j|\mathbf{x})$$

$$= \hat{p}_U^{PCC}(y_i) \tag{4.10}$$

$$E[p_U^h(\hat{y}_j|y_i)] = \frac{1}{|U_i|} \sum_{\mathbf{x} \in U_i} p(y_j|\mathbf{x})$$

and U_i indicates the set of items in U whose true class is y_i. Like for ACC, once the (soft) classifier has been trained and applied to U, the quantity $E[p_U^h(\hat{y}_j)]$ can be observed, and the quantity $E[p_U^h(\hat{y}_j|y_i)]$ can be estimated from L via k-fold cross-validation, which means that we are again in the presence of a system of $|\mathcal{Y}|$ linear equations in $|\mathcal{Y}|$ unknowns, that we can solve by the usual techniques. In the binary case, Equation 4.9 simplifies as

$$E[p_U^h(\hat{\oplus})] = E[p_U^h(\hat{\oplus}|\oplus)] \cdot p_U(\oplus) + E[p_U^h(\hat{\oplus}|\ominus)] \cdot p_U(\ominus) \tag{4.11}$$

from which, similarly to the case of ACC (Equations 4.5 to 4.7), we can derive

$$\hat{p}_U^{PACC}(\oplus) = \frac{\hat{p}_U^{PCC}(\oplus) - E[p_L^h(\hat{\oplus}|\ominus)]}{E[p_L^h(\hat{\oplus}|\oplus)] - E[p_L^h(\hat{\oplus}|\ominus)]} \tag{4.12}$$

where $E[p_L^h(\hat{\oplus}|\oplus)]$ and $E[p_L^h(\hat{\oplus}|\ominus)]$ are the probabilistic counterparts of TPR and FPR in ACC, i.e.,

$$E[p_L^h(\hat{\oplus}|\oplus)] = \frac{1}{|\mathbf{x} \in L : \Phi(\mathbf{x}) = \oplus|} \sum_{\mathbf{x} \in L : \Phi(\mathbf{x}) = \oplus} p(\oplus|\mathbf{x})$$

$$\tag{4.13}$$

$$E[p_L^h(\hat{\oplus}|\ominus)] = \frac{1}{|\mathbf{x} \in L : \Phi(\mathbf{x}) = \ominus|} \sum_{\mathbf{x} \in L : \Phi(\mathbf{x}) = \ominus} p(\oplus|\mathbf{x})$$

PACC was first proposed by Bella et al. (2010). Like PCC, also PACC is dismissed as unsuitable by Forman (2005, 2008), essentially for the same reasons for which he also dismisses PCC and already mentioned in Section 4.2.2.

Like ACC, also PACC can return values for $\hat{p}_U(y_i)$ that fall off the [0,1] range. Again, clipping and rescaling has been used in the literature to deal with these cases; again, applying a softmax (as suggested in Section 4.2.3 for ACC) may prove a better idea.

4.2.5 X, MAX, and Threshold@0.50

The methods we will describe in this section are binary-only quantification methods (i.e., multiclass versions have not been discussed in the literature, and are non-obvious) proposed by Forman (2006, 2008) and arising from a critical analysis of the ACC method.

Assume a binary quantification task with classes $\mathcal{Y} = \{\oplus, \ominus\}$. Equation 4.7 is such that, in principle, $\hat{p}_U^{ACC}(\oplus)$ is undefined when $TPR_L = FPR_L$: however this is not problematic in practice, since a classifier such that TPR = FPR is, as observed by Forman (2005), too bad to arise in real-life situations (since it is usually the case that TPR is higher, or much higher, than FPR). Forman (2008) points out that ACC is very sensitive to the decision threshold of the classifier, which may make $\hat{p}_U^{ACC}(\oplus)$ behave erratically. In particular he points out that, if $\oplus$ is an infrequent class and the classifier is optimised for standard accuracy, the classifier may have a tendency to almost always predict $\ominus$, i.e., to deliver very small values of TPR and FPR. With such small values, the denominator of Equation 4.4 may be highly unstable and very small anyway, thus jeopardising the method. The methods discussed in this section are instances of ACC that use a threshold different from the standard one: in particular, Forman devised these methods with the goal of choosing "a threshold that admits more true positives and many more false positives, yielding worse classifier accuracy but better quantifier accuracy".

One solution proposed by Forman (2006, 2008) is to heuristically set the decision threshold in such a way that $FPR_L = 1 - TPR_L$ (this method is dubbed X) and then use Equation 4.7. The claimed rationale of this heuristics is to avoid the tails of the $FPR_L(t)$ and $1 - TPR_L(t)$ curves, where t is the decision threshold.

An alternative heuristic that Forman (2006, 2008) discusses is to set the decision threshold in such a way that $(TPR_L - FPR_L)$ is maximised (this is dubbed MAX). Here, the rationale is to avoid small values in the denominator of Equation 4.7, with the goal of avoiding the above-mentioned instability in the final values computed by the equation.

Yet another heuristics proposed by Forman (2006, 2008) is to set the decision threshold in such a way that TPR_L is equal to .50 and then use Equation 4.7; this method is dubbed *Threshold@0.50* (T50). The reason why the author proposes this is that such a threshold tends to be good at avoiding the tail of the $1 - TPR_L(t)$ curve.

Forman (2008) argues that it is especially in highly imbalanced datasets that $(\text{TPR}_L - \text{FPR}_L)$ risks being low; the three methods introduced in this section are thus meant to be helpful especially in contexts characterised by high imbalance.

One problem that seems to affect these methods is that, while the thresholds they choose tend to have some desired properties (e.g., avoiding small values in the denominator of Equation 4.7), these properties do not seem correlated to the one property which would seem of interest here, i.e., the fact that the resulting values of TPR_L and FPR_L are accurate estimates of TPR_U and FPR_U.

4.2.6   Median Sweep

An alternative, binary-only quantification method, proposed by Forman (2006, 2008), consists of computing $\hat{p}_U^{\text{ACC}}(\oplus)$ for every decision threshold that gives rise (in k-fold cross-validation) to different TPR_L or FPR_L values, and take the median of all the resulting estimates of $\hat{p}_U^{\text{ACC}}(\oplus)$. This method is dubbed *Median Sweep* (MS), and its rationale lies in the ability of the median to avoid outliers. Following this intuition, Forman (2006, 2008) proposed another variant of this method, called MS2, which computes the median only for cases in which $\text{TPR}_L - \text{FPR}_L > 0.25$.

Again, similarly to what we said about the methods discussed in Section 4.2.5, the problem with this method is that there does not seem to be any *a priori* reason that the median of the estimates of $\hat{p}_U^{\text{ACC}}(\oplus)$ brought about by all possible decision thresholds, while probably not an outlier, is any closer to the true value of $p_U^{\text{ACC}}(\oplus)$ than any of the estimates generated by other "legitimate" methods such as CC or ACC.

4.2.7   The Ratio Estimator

The ACC and PACC methods (Sections 4.2.3, 4.2.4) along with their heuristic spin-offs (Sections 4.2.5 and 4.2.6), have one aspect in common: the solution they propose is based on a specialised version of a general equation, which has the form

$$\hat{p}_U^{\text{RE}}(\oplus) = \frac{E_U[g(\mathbf{x})] - E_L[g(\mathbf{x})|\Phi(\mathbf{x}) = \ominus]}{E_L[g(\mathbf{x})|\Phi(\mathbf{x}) = \oplus] - E_L[g(\mathbf{x})|\Phi(\mathbf{x}) = \ominus]} \qquad (4.14)$$

(for simplicity we here deal with the binary case) where $E_\sigma[x]$ indicates, as usual, the expected value of x in sample σ. Here $g(\mathbf{x})$ is a function of the covariates which can be specialised to obtain Equations 4.7 and 4.12 as

- $g(\mathbf{x}) = \mathbb{1}(h(\mathbf{x}) = \oplus)$ (i.e., the function that is equal to 1 if $h(\mathbf{x}) = \oplus$, and is equal to 0 otherwise) for ACC, i.e., the output of a hard classifier;
- $g(\mathbf{x}) = p(\oplus|\mathbf{x})$ for PACC, i.e., the output of a soft classifier.

This is a key result from Fernandes Vaz et al. (2019), where the authors dub the family of functions described by Equation 4.14 the *ratio estimator* (RE), which is shown, in its entirety, to be Fisher-consistent (a property defined in Section 4.2.3) under prior probability shift. Moreover, the authors prove a Central Limit Theorem (CLT) for RE, allowing the practitioner to approximate the mean squared error (MSE) for the prevalence estimates. Subsequently, they propose a way for selecting the function $g(\mathbf{x})$ based on explicit MSE minimisation, as estimated via the CLT, i.e.,

$$\text{MSE}(\hat{p}_U(\oplus)) \simeq \frac{1}{(\hat{\mu}_\oplus - \hat{\mu}_\ominus)^2 |L|} \left(\frac{\hat{p}_U(\oplus)^2 \hat{s}_\oplus^2}{p_L(\oplus)} + \frac{\hat{p}_U(\ominus)^2 \hat{s}_\ominus^2}{p_L(\ominus)} \right) \tag{4.15}$$

where we have defined the sample moments

$$\hat{\mu}_\oplus = \frac{1}{|\{\mathbf{x} \in L : \Phi(\mathbf{x}) = \oplus\}|} \sum_{\mathbf{x} \in L : \Phi(\mathbf{x}) = \oplus} g(\mathbf{x})$$

$$\hat{s}_\oplus^2 = \frac{1}{|\{\mathbf{x} \in L : \Phi(\mathbf{x}) = \oplus\}|} \sum_{\mathbf{x} \in L : \Phi(\mathbf{x}) = \oplus} (g(\mathbf{x}) - \hat{\mu}_\oplus)^2$$

Here MSE is clearly shown to decrease when the difference between $\hat{\mu}_\oplus$ and $\hat{\mu}_\ominus$ is high. This provides theoretical support for choosing $g(\mathbf{x})$ as the output of a classifier (a function specifically aimed at separating $\oplus$ from $\ominus$), as is the case with ACC and PACC. In general, it is worth noting that a clear characterisation of the relationship between classification and quantification performance remains under-explored in the literature, except for some recent initial results (Tasche, 2021). Additionally, the CLT proved for RE can be exploited to compute confidence intervals (CI) for quantification estimates.[9] CIs for prevalence estimation are discussed more in depth in Section 5.8.

Finally, the authors generalise their results to two novel quantification scenarios. In one scenario, some labels from the target population U are available, thus providing some additional information, not available in typical quantification settings, which can be exploited via a weighted average

$$\hat{p}_U^{\text{AVG}}(\oplus) = w \cdot \hat{p}_U^{\text{RE}}(\oplus) + (1 - w) \cdot \hat{p}_U^{\text{ML}}(\oplus) \tag{4.16}$$

between the ratio estimator $\hat{p}_U^{\text{RE}}(\oplus)$ and a maximum likelihood estimator $\hat{p}_U^{\text{ML}}(\oplus)$ based on available labels from population U (Section 4.1), weighted according to

---

[9] CI are typically defined with respect to a coverage level $(1 - \alpha)$, and define a real-valued interval $[\hat{\Theta}_{lo}, \hat{\Theta}_{hi}] \subseteq [0, 1]$ around an estimation point $\hat{\Theta}$ so that the probability (from a frequentist point of view) for the interval to contain the true value Θ^* is $(1 - \alpha)\%$.

the MSE of each estimator. This setting is related to active learning in data streams, described in Section 5.6.

In a second scenario, we are interested in an estimate of prevalence with finer granularity, dependent on a covariate of interest z. One example of practical interest described by this scenario is the release of an improvement for a given product. In this case one might be interested in verifying the users' reaction to the novelty by segmenting the population of user reviews according to a temporal variable z. A minimal extension of the ratio estimator, i.e.,

$$\hat{p}_U^{\text{RE}}(\oplus|z) = \frac{E_U[g(\mathbf{x})|z] - E_L[g(\mathbf{x})|\ominus]}{E_L[g(\mathbf{x})|\oplus] - E_L[g(\mathbf{x})|\ominus]} \tag{4.17}$$

can solve this task.

4.2.8 Mixture Models

Forman (2005, 2008) proposed a method for quantification based on Mixture Models (MM). MM is yet another binary-only quantification method (i.e., no SLQ extension has surfaced in the literature to date). MM assumes that the cumulative distribution F^U (shorthand for $F^U(s(\mathbf{x}))$) of the scores assigned to data points in U is a mixture

$$F^U = p_U(\oplus) \cdot F_\oplus^U + p_U(\ominus) \cdot F_\ominus^U \tag{4.18}$$

where $F_\oplus^U$ and $F_\ominus^U$ are the cumulative distributions of the scores that the classifier assigns to the positive and to the negative unlabelled examples, respectively, and where $p_U(\oplus)$ and $p_U(\ominus) = (1 - p_U(\oplus))$ are the parameters of this mixture. The MM method consists of estimating $F_\oplus^U$ and $F_\ominus^U$ via k-fold cross-validation on L, and picking as value of $p_U(\oplus)$ the one that generates the best fit between the observed F^U and the mixture. It is worth noting that this approach may work with generic scoring functions $s(\mathbf{x})$ that are not necessarily the output of a soft classifier.

Two variants of this method, called the *Kolmogorov-Smirnov Mixture Model* (MM(KS)) and the *PP-Area Mixture Model* (MM(PP)), are actually defined by Forman (2005), which differ in terms of how the goodness of fit between the left- and the right-hand side of Equation 4.18 is estimated.

Essentially, any method for measuring this goodness of fit can be used in connection with the MM method. Another MM method is HDy, proposed by González-Castro et al. (2013). The difference between HDy and the two previously discussed methods is that in HDy the *Hellinger Distance* (HD, an instance of the class of divergences, that we discussed in Section 3.1) is used to compare two distributions, i.e.,

$$\hat{p}_U^{\text{HDy}}(\oplus) = \text{HDy}(f_{\oplus}^L, f_{\ominus}^L, f^U)$$

$$= \arg\min_{0 \leq \alpha \leq 1} \{\text{HD}(\alpha f_{\oplus}^L + (1-\alpha)f_{\ominus}^L, f^U)\} \tag{4.19}$$

where $f_{\oplus}^L$ and $f_{\ominus}^L$ are the probability density functions of scores (e.g., the output of a soft classifier) for the positive and negative samples of L, respectively, obtained via k-fold cross-validation on L, and f^U is the distribution of scores obtained for U by the classifier trained on L. Notice that $f_{\oplus}^L$, $f_{\ominus}^L$, and f^U are empirically approximated with histograms.

Maletzke et al. (2019) proposed the *Distribution y-Similarity* (DyS) framework, that generalises the HDy approach by considering the dissimilarity function DS as a parameter of the model. A dissimilarity function compares two probability distributions, i.e., the same process of HDy, but uses a different distance function. In this case, the authors approximate probability distributions with histograms, and test a variety of distance functions (Maletzke et al., 2019, Table 1), i.e., Squared Euclidean (SEc), Manhattan (MH), Probabilistic Symmetric (PS), Topsøe (TD), Jensen Difference (JD), Taneja (TN), Hellinger (HD), Dice (DC), Jaccard (JC), Chebyshev (CB), Inner Product (IP), Kumar-Hassebrook (HB), Cosine (CS), and Harmonic Mean (HM). The authors also propose distance functions that do not operate on distributions but directly compare the scores assigned to samples, i.e., Mixable Kolmogorov-Smirnov (MKS) and Sample Ordinal Distance (SORD)[10]. The DyS framework is defined as

$$\hat{p}_U^{\text{DyS}}(\oplus) = \text{DyS}(f_{\oplus}^L, f_{\ominus}^L, f^U)$$

$$= \arg\min_{0 \leq \alpha \leq 1} \{\text{DS}(\alpha f_{\oplus}^L + (1-\alpha)f_{\ominus}^L, f^U)\} \tag{4.20}$$

where $f_{\oplus}^L$, $f_{\ominus}^L$, and f^U are the same probability distributions that appear in Equation 4.19.

Experiments in Maletzke et al. (2019) indicate that the Topsøe distance performs better than all the other compared distance measures. The Topsøe distance is a symmetric version of the Kullback-Leibler divergence (Johnson and Sinanovic, 2001), and is defined as

$$\text{TD}(f, g) = \text{KLD}(f, m) + \text{KLD}(g, m) \tag{4.21}$$

where $m = \frac{1}{2}(f + g)$.

---

[10] For MKS and SORD, Maletzke et al. (2019) presents a specific implementation of the Equation 4.20 that computes the distance function from the classification scores rather than from probability distributions.

Moreira dos Reis et al. (2018b) explore the use of HDy in a *recurrent contexts scenario*, i.e., they assume that the distribution of the data may change only among a limited set of possible distributions. They assume the availability of training data for all the possible contexts L_{C_i} for $i \in \{1, 2, .., |C|\}$, each one representing a possible distribution. They propose two extensions of the HDy method, i.e.,

- *Single Most Relevant HDy* (SMR-HDy). This method applies HDy (Equation 4.19) to each context L_{C_i}, selecting the context L_{C_m} that minimises the Hellinger distance, and returns the prevalence estimate associated to that context, i.e.,

$$C_m = \arg\min_{i \in C} \text{HDy}(f_{\oplus}^{L_{C_i}}, f_{\ominus}^{L_{C_i}}, f^U) \tag{4.22}$$

$$\hat{p}_U^{\text{SMR−HDy}}(\oplus) = \text{HDy}(f_{\oplus}^{L_{C_m}}, f_{\ominus}^{L_{C_m}}, f^U) \tag{4.23}$$

- *Crossed-Opinions HDy* (XO-HDy), in which the data in every context is split into two parts, train Tr_{C_i} and validation Va_{C_i}, resulting in more complex procedure for the selection of the most likely context C_m. Specifically, each training set is used to learn a classifier, which is then applied to every validation set, producing $|C|^2$ classifications. The distribution $f^{L_{ij}}$ for each of such classifications is computed. The distribution of U is compared with all the distributions $f^{L_{ij}}$ to find the most plausible context for unlabelled data, i.e.,

$$C_m = \arg\min_{j \in C} \frac{1}{|C|} \sum_{i \in C} \text{HD}(\alpha_{ij} f_{\oplus}^{L_{ij}} + (1 - \alpha_{ij}) f_{\ominus}^{L_{ij}}, f^{U_i}) \tag{4.24}$$

where $f^{L_{ij}}$ is the distribution of the scores obtained by a classifier trained in context C_i on the validation set from context C_j, f^{U_i} is the distribution obtained by the same classifier on U, and $\alpha_{ij} = \text{HDy}(f_{\oplus}^{L_{ij}}, f_{\ominus}^{L_{ij}}, f^{U_i})$. Finally, XO-HDy returns the prevalence estimate $\hat{p}_U^{\text{HDy}, L_{C_m}}(\oplus)$ associated to that context. Notice that, after selecting the most likely context for U, prevalence estimate can be provided by quantification methods other than HDy.

Also grounded in mixture models, the method *Gain-Some-Lose-Some* (GSLS) Denham et al. (2021) was proposed as a means to counter the effects of dataset shift in class prevalence estimation. The authors argue that GSLS is designed to deal with forms of dataset shift other than prior probability shift.

The method assumes that the observed probability distribution $f^L(\mathbf{x})$ and the target distribution $f^U(\mathbf{x})$, hereafter shortened to f^L and f^U, are related by means of an intermediate distribution f^R. This intermediate distribution is indicated with an R standing for "remaining distribution", since the framework assumes that the source distribution f^L can be composed as a mixture of the f^R distribution and a *loss* distribution f^- (i.e., that f^R is *what remains* after *losing* (subtracting) a distribution f^- from f^L) and f^U can be composed as a mixture of the f^R

distribution and a *gain* distribution f^+ (i.e., the target distribution f^U is obtained from f^R by adding a *gain* distribution f^+). That is,

$$f^L = w^- f^- + (1 - w^-) f^R \tag{4.25}$$

$$f^U = w^+ f^+ + (1 - w^+) f^R \tag{4.26}$$

where w^- and w^+ are the weights of the mixtures, i.e., the amount of loss and gain, respectively. Note that the distributions f^- and f^+ refer to subpopulations, *lose* and *gain* respectively, and have nothing to do with the classes of a binary problem; indeed, GSLS is formulated for the general multiclass setting.

Without further assumptions, there are infinitely many ways for choosing distributions f^-, f^+, and f^R, and weights w^- and w^+. GSLS thus makes some simplifications and imposes some constraints to render the problem tractable. One such simplification consists of modelling, as other methods like HDy do, the distributions as b-bin histograms of the outputs produced by a probabilistic classifier. For doing so, and in order to avoid overfitting, the classifier is trained on separate source data, and the histogram is computed on held-out validation source data. Then, the unknowns f^R, f^-, f^+, w^-, and w^+ are searched by solving an optimisation problem that attempts to minimise the degree of (gain and loss) shift. This is akin to minimising $w^- + w^+$ and constraining the bins of the histogram for f^R to lie between those of f^L and f^U.

Once all distributions and mixture weights have been fixed, GSLS uses knowledge from the source distribution to compute quantification predictions, along with their corresponding confidence intervals, for the classes of interest. GSLS computes $p_U(\oplus)$ via maximum likelihood, for which a number of assumptions are needed. The most important one comes down to assuming the proportion of target samples belonging to $\oplus$ to follow a binomial distribution $B(|U|, p_U(\oplus))$ scaled by $\frac{1}{|U|}$, and establishing a relation between the unknown $p_U(\oplus)$ and the (already known) factors of the mixture model. Further assumptions on the underlying distributions allow GSLS to express this probability in terms of other parametric distributions; the complete mathematical derivation is explained in Denham et al. (2021).

4.2.9 Expectation Maximisation for Quantification

All the methods discussed so far have an *inductive* nature, since the quantification model is trained exclusively on the training set. Some quantification methods proposed in the literature have instead a *transductive* nature (see e.g., Joachims, 1999), i.e., they are trained by also looking at certain characteristics of the unlabelled examples they need to issue predictions for (although not at their labels). Because of this, the model generated may fit the designated test set better than the models generated via inductive methods, but is less general, since it is especially tailored to the very set of unlabelled items used in the training phase, and may underperform when applied to different sets of unlabelled items. The *Saerens-Latinne-Decaestecker* (SLD) algorithm, proposed by Saerens et al. (2002), has a transductive component, since it applies a transductive correction to the test predictions (issued by an inductive classifier).

SLD is an instance of Expectation Maximisation (Dempster et al., 1977), the well-known iterative algorithm for finding maximum-likelihood estimates of parameters (in our case: the class prevalence values) for models that depend on unobserved variables (in our case: the class labels). Essentially, SLD (see Algorithm 1) incrementally updates (Line 10) the posterior probabilities by using the class prevalence values computed in the last step of the iteration, and updates (Line 14) the class prevalence values by using the posterior probabilities computed in the last step of the iteration, in a mutually recursive fashion.

Input : Class prevalence values $p_L(y)$ on L, for all $y \in \mathcal{Y}$;
　　　　　 Posterior probabilities $p(y|\mathbf{x})$, for all $y \in \mathcal{Y}$ and for all $\mathbf{x} \in U$;
Output: Estimates $\hat{p}_U(y)$ of class prevalence values on U;

```
/* Initialisation                                              */
1  s ← 0;
2  for y ∈ 𝒴 do
3  │    p̂_U^(s)(y) ← p_L(y);
4  │    for x ∈ U do
5  │    │    p^(s)(y|x) ← p(y|x);
6  │    end
7  end
```

```
/* Main Iteration Cycle                                        */
8  while stopping condition = false do
9  │    s ← s + 1;
10 │    for y ∈ 𝒴 do
11 │    │    for x ∈ U do
```

$$12 \qquad p^{(s)}(y|\mathbf{x}) \leftarrow \frac{\dfrac{\hat{p}_U^{(s-1)}(y)}{\hat{p}_U^{(0)}(y)} \cdot p^{(0)}(y|\mathbf{x})}{\sum_{y \in \mathcal{Y}} \dfrac{\hat{p}_U^{(s-1)}(y)}{\hat{p}_U^{(0)}(y)} \cdot p^{(0)}(y|\mathbf{x})}$$

```
13 │    │    end
```

$$14 \qquad \hat{p}_U^{(s)}(y) \leftarrow \frac{1}{|U|} \sum_{\mathbf{x} \in U} p^{(s)}(y|\mathbf{x})$$

```
15 │    end
16 end
```

```
/* Generate output                                             */
17 for y ∈ 𝒴 do
18 │    p̂_U(y) ← p̂_U^(s)(y)
19 end
```

Algorithm 1: The SLD algorithm (Saerens et al., 2002).

Like ACC (Section 4.2.3), SLD was proven to be *Fisher-consistent under prior probability shift* (Tasche, 2017). In the same work, the author provides a counterexample of dataset shift under which SLD loses Fisher consistency.

Alaíz-Rodríguez et al. (2011) propose an extension of SLD, based on the assumption that each class can be decomposed into several subclasses and that the change in the prevalence of the class is actually determined by the change in the prevalence of its subclasses[11]. The method that Alaíz-Rodríguez et al. (2011) propose consists of two main steps. The first step consists of estimating the number of subclasses and their prior probabilities. To do so, an iterative method called *Posteriori Probability Model Selection* (PPMS) Arribas and Cid-Sueiro (2005) is applied to L. PPMS applies pruning, splitting, and merging criteria, to dynamically choose the optimal number of subclasses of each class during training. The output is not only the number of subclasses per class, but also their prior probabilities and the posterior probabilities of each item, as computed by a two-layered feedforward network called *Generalised Softmax Perceptron* (GPS) (Guerrero-Curieses et al., 2005). The second step applies an extension of SLD that jointly adjusts for the class and subclass probabilities.

Re-estimating class prevalence values at subclass level was empirically shown to yield improved results when compared to SLD in the experiments of Alaíz-Rodríguez et al. (2011). One interesting experiment showed that (artificially) introducing concept shift at subclass level (within classes), yet leaving the prior probabilities unchanged at the class level, might cause the application of SLD (at class level) to be detrimental with respect to not performing re-estimation at all.

4.2.10 Class Distribution Estimation

Xue and Weiss (2009) propose a similar procedure dubbed *Class Distribution Estimation Iterate* (CDE-Iterate). This procedure is primarily aimed at improving classification accuracy, while the improvement of quantification estimates (i.e., class priors) is considered by the authors as an accessory step toward the main goal.

CDE-Iterate employs *cost-sensitive learning*, by assigning different values to the cost of false negatives (c_{FN}), and false positives (c_{FP}). The ratio between the two costs is kept proportional to a value associated with the shift in prior probabilities, by enforcing

$$c_{FP} = \frac{p_L(\oplus)\,\hat{p}_U(\ominus)}{p_L(\ominus)\,\hat{p}_U(\oplus)} c_{FN} \qquad (4.27)$$

---

[11] The problem setting they address is single-label, both at class and subclass levels (i.e., data items labelled as y_i belong to strictly one of the j subclasses of y_i).

This probabilities shift is defined as the ratio between positive-to-negative rate in L and positive-to-negative rate in U; the former is readily available, whereas the latter is estimated via Classify and Count, which is also the quantification method used to determine quantification estimates.

The algorithm, iterative and transductive in nature, starts training a cost-sensitive hard classifier h on L, using $c_{FP} = c_{FN} = 1$. The value c_{FN} does not change through the execution of the method. The main iteration of the algorithm consists of the following steps:

1. Run a hard classifier h on U, thus obtaining $\hat{p}_U(\oplus)$, $\hat{p}_U(\ominus)$.
2. Update c_{FP} according to Equation 4.27.
3. Retrain h on L in a cost-sensitive fashion with new values for c_{FP}, c_{FN}.

A key disadvantage with respect to SLD is the need to retrain a cost-sensitive classifier h at each iteration, slightly compensated by the possibility of wrapping CDE-Iterate around hard classifiers, without requiring h to output posterior probabilities. A further disadvantage is the lack of Fisher consistency under prior-probability shift.

4.2.11 Ensemble Methods for Quantification

Attempts have been made at characterising the applicability of ensemble techniques to problems of binary quantification. This paradigm was proposed in early quantification work (Forman, 2006, 2008), which focuses on the choice of an optimal threshold for a classifier that would allow a good estimate for true positive rate and false positive rate. An approach dubbed *Median Sweep* (Section 4.2.6) is proposed, which considers different classifier thresholds each yielding a different estimate of class prevalence via Adjusted Classify and Count. The estimates are aggregated by computing their median, which is regarded as the final quantification result.

In more recent years, a line of work has emerged in quantification literature, solely focused on ensembles (Pérez-Gállego et al., 2017, 2019). The key idea for this paradigm is training multiple quantifiers introducing diversity in the skew level of the training set employed for each model. At testing time, the outputs from each model (or a subset thereof) are suitably aggregated into a single prevalence estimate. In its simplest form, the algorithm can be conceptually divided into 3 steps:

1. Sample generation is carried out by sampling an unlabelled prevalence p_i at random from $[0, 1]$. For each p_i, a training set L_i is then generated enforcing a prior $p_{L_i}(y) = p_i$ for positive class via random sampling with replacement.
2. Model training is performed on each sample L_i generated under the procedure described above. Three quantification algorithms are considered for this step, namely *Classify and Count* (Section 4.2.1), *Adjusted Classify and Count* (Section 4.2.3), and HDy (Section 4.2.8).

3. Output aggregation: the final estimate for class prevalence of the unlabelled set U is computed as the arithmetic mean of the outputs from all models from the ensemble.

Pérez-Gállego et al. (2019) expand on the above work by considering different combination techniques for output aggregation (Step 3). The strategy employed consists of discarding half of the learnt models, thereafter averaging the output of the remaining ones. If model selection is carried out considering only labelled data, the resulting procedure is dubbed *static*, as opposed to a *dynamic* approach whereby the choice also takes into account the unlabelled dataset U. Two *static* methods are proposed for the selection of the strongest models:

- An accuracy-based approach ranks models according to the Mean Squared Error they exhibit when tested on every sample L_i of the training set generated during Step 1.
- An approach inspired by the algorithms described in Section 4.2.5, and aimed at producing stable adjustments to class prevalence estimates, e.g., by maximising the denominator from Equation 4.5.

Two more criteria are proposed that embrace a *dynamic* approach:

- *Training prevalence* runs all quantifiers on U and ranks them according to the difference between the mean estimated prevalence for U and the prevalence in the training set L_i used for training them.
- *Distribution similarity*, inspired by the work described in Section 4.4.4, compares the distribution of posteriors $p(y|x)$ between L_i and U, ranking each quantifier based on the Hellinger distance computed on histograms.

After ranking models based on *static* or *dynamic* criteria, the top half is selected and the respective estimates are averaged, thus yielding the final estimate of class prevalence.

4.2.12 QuaNet

A first attempt at using a deep neural network for text quantification is presented in Esuli et al. (2018). The network takes as input the classification scores for each document in the sample to be quantified produced by document classifier, a document embedding for each document in the sample, and the output of an ensemble of quantification methods.

The QuaNet neural network is composed of two main parts (see Figure 4.1). Given a set of documents on which to perform the quantification, a first part of the network takes as input the sequence of document embeddings sorted by the classification score assigned by a document classifier. This part of the network is composed of an LSTM that, by observing how the content of the documents varies (as represented in the document embeddings) in relation to the classification scores,

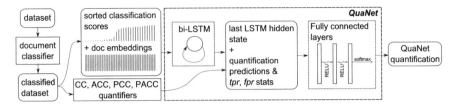

Fig. 4.1 Architecture of the QuaNet network, from Esuli et al. (2018).

learns to output a "quantification embedding" which captures the composition of the whole set of documents to be quantified.

The second part of the network takes as input such quantification embedding vector as well as the estimated prevalence values from the ensemble of quantification methods, i.e., those described by Equations 4.2, 4.3, 4.4, 4.9 (i.e., $\hat{p}_U^{CC}(\oplus)$, $\hat{p}_U^{PCC}(\oplus)$, $p_U^h(\hat{\oplus}_i)$, $E[p_U^h(\hat{\oplus}_i)]$), and other statistics on the underlying classifier (the TPR, FPR, TNR and FNR estimates, see Section 4.2.3). All these values are processed through a set of fully connected layers that output the quantification prediction. Given a training set of labelled documents, the training examples for QuaNet are samples of the training set sampled so as to cover all the possible prevalence values.

The rationale behind QuaNet is that the network learns to select, combine and correct the information coming from a committee of quantification methods, as a function of an abstract representation of the content of the set of documents to be quantified in order to produce a more accurate quantification across all the spectrum of possible prevalence values, whereas each single method maybe more accurate on some ranges than on others. The QuaNet network can in principle work with any classifier, and also the embeddings can have different origin and form, e.g., they can be either traditional bag-of-word sparse representations or dense representations produced by a language model or as the by-product of a classification NN (as done in Esuli et al. (2018)). Similarly, the committee of quantification methods given in input can be varied, with experiments from the original authors confirming the intuition that the richer the committee is, the better the results are.

4.3 Aggregative Methods Based on Special-Purpose Learners

To date, most proposed methods explicitly addressed to quantification (Barranquero et al., 2013; Bella et al., 2010; Forman, 2005, 2006, 2008; Forman et al., 2006; Hopkins and King, 2010; Xue and Weiss, 2009) employ general-purpose supervised learning methods, i.e., address quantification by elaborating on the results returned by a general-purpose classifier. A different stance is taken by the works in this section, which propose the use of learning algorithms explicitly designed with quantification in mind. Said methods propose special optimisation criteria, which are devised to bring about good quantification performance under simple aggregation (i.e., Classify and Count). Tasche (2016) provides an interesting theoretical analysis on this approach, which he dubs *quantification without adjustment*, highlighting some inherent limitations.

4.3.1 Methods Based on Explicit Loss Minimisation

In a position paper, Esuli and Sebastiani (2010b) suggest the use of an *explicit loss minimisation* approach to quantification, based on the idea of using a learning algorithm that is "aware" of the measure (a.k.a. "loss") used for evaluating quantification error, i.e., a learning algorithm that explicitly minimises that measure, whichever it may be. This is an implicit answer to the methods discussed in Section 4.2.5, which all attempt to address the undesired side effects of using learning algorithms that minimise Hamming loss (i.e., "vanilla" classification error), or proxies thereof.

The idea of using classifier-training algorithms capable of directly minimising the measure used for evaluating error is well-established in supervised learning. However, in the case of quantification, following this route is non-trivial, because the functions used for evaluating quantification (see Section 3) are inherently *nonlinear*, i.e., are such that the error on the set of unlabelled items may not be formulated as a linear combination of the error incurred by each unlabelled example. The reason for this inherent non-linearity is that, how the error on an individual unlabelled item impacts on the the error on the set of unlabelled items depends on how the other unlabelled items have been classified. For instance, if in the other unlabelled items there are more false positives than false negatives, an additional false negative is actually *beneficial* to overall quantification error, because of the mutual compensation effect between FP and FN mentioned in Section 1.2. As a result, a measure of quantification error is inherently nonlinear, and should thus be *multivariate*, i.e., take in consideration all the unlabelled items at once.

The assumption that the error on the set of unlabelled items may be formulated as a linear combination of the error incurred by each unlabelled example (as indeed happens for many common error measures – e.g., Hamming distance) underlies most existing learners, which are thus suboptimal for tackling quantification. In order to sidestep this problem, Esuli and Sebastiani (2010b) suggest the use of the *SVM for Multivariate Performance Measures* (SVM$_{perf}$) learning algorithm proposed by Joachims (2005). SVM$_{perf}$ is a "structured output" learning algorithm of the Support Vector Machine family that can generate classifiers optimised for any nonlinear, multivariate loss function that can be computed from a contingency table (as all the measures discussed in Section 3 are). Esuli and Sebastiani (2014, 2015) implement and test the idea, adopting KLD or NKLD as the loss function to be minimised; the SVM(KLD) and SVM(NKLD) methods consist of adopting plain Classify and Count using a classifier generated by SVM$_{perf}$ as instantiated with the KLD or NKLD loss measures.

Barranquero et al. (2015) follow a very similar route, but instead of minimising a "pure" quantification loss they minimise (also via SVM$_{\text{perf}}$) the Q measure, a combination of a classification loss function M^c and a quantification loss function M^q obtained (by mimicking the F_β measure (van Rijsbergen, 1979)) as the harmonic mean between M^c and M^q, i.e.,

$$Q_\beta^{M^c, M^q} = (1 + \beta^2) \frac{M^c \cdot M^q}{\beta^2 \cdot M^c + M^q} \tag{4.28}$$

The rationale of minimising the Q measure is that, by doing so, the authors attempt to learn a good quantifier that is also a good classifier, the underlying idea being that a system that delivers good quantification accuracy but bad classification accuracy is not a trustworthy quantifier. (This idea will be discussed again in Section 4.3.2). In their experiments, Barranquero et al. (2015) use $(1-\text{recall})$ and NAE as the classification and quantification loss functions M^c and M^q in Equation 4.28.

As an alternative to the use of SVM$_{\text{perf}}$ for minimising quantification loss measures, Kar et al. (2016) propose to tackle the explicit minimisation of KLD and other quantification loss measures via an online stochastic optimisation algorithm (*NEsted priMal-dual StochastIc updateS* – NEMSIS) that they devise. NEMSIS is an algorithm for online stochastic optimisation of nested concave functions, i.e., concave functions of functions that are themselves concave; Kar et al. (2016) show that $-$KLD (the negation of KLD) is indeed nested concave, which means that it lends itself to optimisation by means of NEMSIS. Via a similar process, the authors propose online stochastic optimisation algorithms that can deal with several other evaluation measures for quantification, including Barranquero et al.'s (2015) Q measure and variants thereof. Following the same line of research, Sanya et al. (2018) present a family of algorithms that can directly train deep neural networks, and other methods that generate nonlinear classifiers, to optimise quantification loss functions such as KLD.

4.3.2 Quantification Trees and Quantification Forests

Another work that proposes the use of learning technology specially designed for quantification is the one by Milli et al. (2013). This work customises decision trees to deal with quantification, thereby yielding what the authors call *quantification trees*. Like all decision trees (see e.g., Duda et al., 2001, §8.2 for an introduction), quantification trees are built by recursively selecting the best feature for splitting the training data, until a stopping condition is verified. Essentially, a quantification tree is a kind of decision tree in which both (a) the splitting criterion and (b) the stopping condition are informed by measures of quantification accuracy. The authors propose two methods for training a quantification tree, which differ in terms of how step (a) is tackled.

In the first method (called *classification error balancing*), the |FP − FN| measure (a proxy of absolute error) is used for evaluating the quality of a split[12]. For instance, if nodes in the tree check for the presence or the absence of a feature in the unlabelled item (as in binary decisions trees), during the training phase the chosen feature on which to split is the one that minimises the absolute difference between the number of false positives and the number of false negatives resulting from the split.

In the second method (called *Classification-quantification balancing*), the quality of a split is evaluated by the function

$$\text{MOM}(p, \hat{p}) = |\text{FP}^2 - \text{FN}^2| = (\text{FP} + \text{FN})(|\text{FP} - \text{FN}|) \qquad (4.29)$$

The rationale of MOM (which stands for *multi-objective measure*) is that (FP + FN) is a measure of *classification* error, while |FP − FN| is a measure of *quantification* error, which means that by minimising their product one attempts to generate low values of both quantities at the same time. The underlying intuition is that it is difficult to trust a quantifier if it is not also a good classifier, and by attempting to simultaneously maximise both classification and quantification accuracy we thus strive to obtain good and *trustworthy* quantifiers. (Such an attempt is also the rationale of Barranquero et al. (2015), a work that we have discussed in Section 4.3.1). The reader may have noticed that we had not mentioned MOM in Section 3.1, i.e, when discussing evaluation measures for quantification. The reason is that MOM, while a reasonable measure for a learner to optimise, is not a reasonable measure for evaluating the results of a quantifier, because it does not evaluate quantification error but a combination of quantification error and classification error.

Concerning step (b), Milli et al. (2013) stop growing the tree when no possible split would bring about an improvement in the chosen measure of quality (a measure which differs, of course, depending on which of the two methods above is used). Milli et al. (2013) take this approach further by proposing *Quantification Forests* (QFs). Essentially, a quantification forest is a "decision forest" (also known as "random forest" – see Criminisi et al. (2011) for an introduction) of quantification trees: a set of quantification trees is generated (each by restricting the training set to k_1 randomly chosen training documents and k_2 randomly chosen features), and the average of the prevalence estimates for class ⊕ is chosen as the final prevalence estimate $\hat{p}(⊕)$.

Note that quantification trees and quantification forests can be used either in their pure form (i.e., using a CC-style method as in Section 4.2.1) or, as Milli et al. (2013) indeed do, by applying the ACC-style correction of Section 4.2.3 to the class prevalence estimates they generate.

---

[12] The authors also mention the possibility of directly using KLD as a loss (i.e., as a measure of quality of the split), but do not present experiments on this.

4.4 Non-Aggregative Methods

So far, we have discussed methods that work by aggregating the individual decisions that a (hard or soft) classifier takes for each and every unlabelled item, and possibly performing some post-processing. However, this is not the only possible route to quantification, and systems that estimate class prevalence values without generating binary decisions or posterior probabilities for the individual items as an intermediate step, can be conceived. Indeed, this route has a theoretical justification in the so-called *Vapnik's principle*, that states (Vapnik, 1998)

> "If you possess a restricted amount of information for solving some problem, try to solve the problem directly and never solve a more general problem as an intermediate step. It is possible that the available information is sufficient for a direct solution but is insufficient for solving a more general intermediate problem."

This principle is directly applicable to quantification, since classification is a more general problem than quantification: in fact, if we have a (hard or soft) classifier we also have a quantifier, since in order to estimate class prevalence values we only need to apply (Probabilistic or non-) Classify and Count, but if we have a quantifier (i.e., an estimator of class prevalence values) this does not mean that we have a classifier. Vapnik's principle suggests that the information in the training set might be sufficient for solving quantification directly but not for solving it *indirectly*, i.e., for training a classifier that classifies the individual documents as an intermediate step. Non-aggregative quantifiers are thus the ones that more closely follow the spirit of Vapnik's principle; this section is devoted to such non-aggregative methods for learning to quantify.

4.4.1 The READNME Method

The method proposed by King and Lu (2008), later named READNME and popularised by Hopkins and King (2010), is a text quantification method based on the idea of estimating class prevalence values directly via equation

$$p(\mathbf{x}_i) = \sum_{y_j \in \mathcal{Y}} p(\mathbf{x}_i | y_j) p(y_j) \tag{4.30}$$

where $p(\mathbf{x}_i)$ represents the probability that a document drawn at random from U has $\mathbf{x}_i$ as its vectorial representation. The problem is framed, using matrix notation, as

$$p_U(\mathcal{X}) = p_U(\mathcal{X}|\mathcal{Y}) p_U(\mathcal{Y}) \tag{4.31}$$

where $p_U(\mathcal{X})$ is a $2^K \times 1$ vector whose elements are the probability of each possible variate (binary vector) of K features, $p_U(\mathcal{X}|\mathcal{Y})$ is a $2^K \times |\mathcal{Y}|$ matrix where the j-th column has the class-conditional probabilities of all possible variates, and $p_U(\mathcal{Y})$ is the $|\mathcal{Y}| \times 1$ class prevalence array of interest; the solution to this equation can either be achieved by standard constrained least squares as

$$\hat{p}_U(\mathcal{Y}) = (p_U(\mathcal{X}|\mathcal{Y})^\top p_U(\mathcal{X}|\mathcal{Y}))^{-1} p_U(\mathcal{X}|\mathcal{Y})^\top p_U(\mathcal{X}) \qquad (4.32)$$

and then replacing $p_U(\mathcal{X}|\mathcal{Y})$ with $p_L(\mathcal{X}|\mathcal{Y})$ under the assumption that the class-conditional probabilities $p(\mathbf{x}_i|y_j)$ remain invariant between the training and unlabelled data[13].

Of course, the problem is that in high-dimensional spaces (such as in the standard "bag-of-words" representation used in text-related applications), the dimension 2^K affecting $p_L(\mathcal{X}|\mathcal{Y})$ and $p_U(\mathcal{X})$ rapidly explodes, causing the method to become computationally intractable. To solve this issue, Hopkins and King (2010) applied bagging, i.e., repeatedly taking random subsets of features ("between approximately 5 and 25 words" long) and estimating $p(y_j)$ as the average of several runs. They also applied bootstrapping to re-sample matrix rows and estimate the method variance and thus deriving confidence intervals of the estimation.

Hopkins and King (2010) perform a small-scale experimentation (using 4 datasets, with sizes ranging from 462 to 4303 documents) in which their method is shown to outperform four baselines, each consisting of the CC method as applied to an SVM with a different kernel (linear, radial, polynomial, sigmoid). However, no details are given as to the number of subsets of the vectorial representation and size of these subsets used in these experiments. One drawback of this method is that it depends on several hyperparameters, e.g., the number of subsets of the vectorial representation, and the size of these subsets. Finding optimal values for these hyperparameters may thus require extensive cross-validation.

4.4.2   The iSA Method

Although README (Hopkins and King, 2010) already counters some computational issues presented in the original version by King and Lu (2008), it still demands a considerable amount of computational power, mainly due to the application of bagging. Ceron et al. (2016) proposed a variant called iSA (standing for *integrated Sentiment Analysis*) that gets rid of the bagging approach by first applying a

---

[13] King and Lu (2008) argued this assumption to hold whenever the "data generation process" falls within the type $\mathcal{Y} \to \mathcal{X}$, that is, when the class variable turns to condition the distribution of the variates in the feature space $\mathcal{X}$. While this might stand true in their applicative scenario (verbal autopsies), where the causes of death might determine the symptoms, said assumption might not hold in general, nor be easily verifiable in practice. All other things being equal, and for reasons discussed in Section 1.5, we prefer not to stick to any dichotomy of quantification methods built on top of beliefs about data causality (thus embracing *data generation* considerations, *temporal* dependencies, or *intrinsic/extrinsic* judgements about labels).

series of transformations to the data instances and then directly solving (once) for Equation 4.32. The main transformation consists of artificially augmenting the number of instances by replacing each original data with simpler versions of it. Concretely, iSA replaces each n-dimensional document representation (within a bag-of-words model) with its $b = \lceil n/l \rceil$ (non-overlapping) chunks of length l (with l a parameter to be specified by the user).[14]

4.4.3 The README2 Method

Jerzak et al. (2022) proposed README2 aiming at improving the performance of the original README system by Hopkins and King (2010). This improved version attempts to counter three situations that could degrade the performance of the original method, and that the authors identified as (i) *semantic change*, concerning the differences in meaning of language used across L and U, which can in turn be *emergent* (some terminology appears exclusively in U) or *vanishing* (some terminology appears exclusively in L); (ii) *lack of textual discrimination*, regarding those categories that are hardly distinguishable by the textual features; and (iii) *proportion divergence*, which is analogous to the prior probability shift between L and U.

README2 introduces two main novelties with respect to the former version, seeking to better represent the meaning of text. The first one consists of moving away from the sparse representation of the feature space and the subsampling procedure in favour of a dense representation based on word embeddings. The second one consists of improving the feature discrimination by learning a (feed-forward) neural transformation of the resulting matrix which is optimised for quantification. This transformation is formalised as an optimisation problem seeking to satisfy two desirable criteria for the new representation: *category distinctiveness* (the new features brings about more distant class-conditional means across categories), and *feature distinctiveness* (the rows of the transformed matrix present low correlation of one another). Both criteria are implemented as two different loss functions which define, as a weighted sum, the objective loss function to minimise.

Additionally, the aforementioned "vanishing" discourse effect is mitigated by subsampling L in a way that its term distribution gets closer to that in U. Jerzak et al. (2022) observed that selective pruning of L indirectly helped to reduce the "proportion divergence" with respect to U.

---

[14] Actually of length $(l + 1)$, since a positional character informing of the chunk's order in the original sequence is added.

4.4.4 The HDx Method

González-Castro et al. (2013) propose a quantification method for binary problems based on distributional divergence as measured via the Hellinger Distance (HD). The method, referred to as HDx, applies to scenarios in which $p(\mathbf{x}|\oplus)$ is assumed to be fixed but $p(\oplus)$ may vary.

This method is closely related to HDy and other mixture models (Section 4.2.8), with the difference of considering probability distributions $f(\mathbf{x})$ over the multidimensional input domain $\mathcal{X}$, instead of distributions $f(s(\mathbf{x}))$ over single-dimensional scores computed from the input. The rationale is to measure the similarity between the unlabelled distribution and a *validation distribution*, which is generated from the training distribution at a controlled prevalence. HDx iteratively varies this prevalence at small steps ranging from 0 to 1 and seeks the prevalence that maximises the match with the unlabelled distribution as a Mixture Model.

HDx measures the distributional divergence between input data $\mathbf{x}$, as represented in a feature space (e.g., tfidf values) between two distributions f and g via the HD, defined as

$$\mathrm{HD}(f, g) = \sqrt{\int \left(\sqrt{f(\mathbf{x})} - \sqrt{g(\mathbf{x})}\right)^2 dx} \qquad (4.33)$$

The method they propose actually computes this divergence by integrating the HD between each feature distribution independently, which is discretised using bins. The integral is approximated by summing over the bins, i.e.,

$$\mathrm{HD}(V, U) = \frac{1}{n} \sum_{f=1}^{n} \sqrt{\sum_{i=1}^{b} \left(\sqrt{\frac{|V_{fi}|}{|V|}} - \sqrt{\frac{|U_{fi}|}{|U|}}\right)^2} \qquad (4.34)$$

where V is the validation sample, $|V_{fi}|$ is the number of times the feature f appears in the bin i, n is the number of features (e.g., distinct terms), and b is the number of bins. The method thus consist of returning

$$\alpha^* = \arg \min_{\alpha \in [0,1]} \mathrm{HD}(V^\alpha, U) \qquad (4.35)$$

with α^* the prevalence. Actually, V^α is created neither by over- nor by undersampling L, but is instead constructed as a mixture of the class-conditional distributions parameterised with the desired prevalence α, i.e.,

$$V^\alpha(\mathbf{x}) = \alpha \cdot p(\mathbf{x}|\oplus) + (1 - \alpha) \cdot p(\mathbf{x}|\ominus) \qquad (4.36)$$

Since the number of bins b might have a significant impact in the calculation, one typically returns the median of the distribution of the best α's found for a range of b's (typical values are $b \in [10, 20, 30, \ldots, 110]$).

The same authors also propose HDy, previously discussed in Section 4.2.8, which, contrarily to HDx, measures the divergence in a single-dimensional space, which represents the codomain of a soft classifier $s(\mathbf{x})$. The fact that HDy relies on a soft classifier to model $p(\oplus|\mathbf{x})$ precludes it from being considered a pure non-aggregative method. Notice HDy significantly outperforms HDx in the experimental evaluation conducted in González-Castro et al. (2013).

4.4.5 The MMD-RKHS Method

Iyer et al. (2014) formulate the quantification problem in terms of minimising the Maximum Mean Discrepancy measure in a Reproducing Kernel Hilbert Space (MMD-RKHS). They prove some error bounds on the application of MMD to quantification and use such theoretical results to define a kernel learning method that minimises the MMD between the observed $p_U(\mathbf{x})$ and $\sum_y p_L(\mathbf{x}|y)\Theta_y$, under the assumption that $p_U(\mathbf{x}|y) = p_L(\mathbf{x}|y)$, where Θ_y are the unknown prevalence values to be estimated. They compare MMD-RKHS against the method of du Plessis and Sugiyama (2012), obtaining similar or slightly better results.

4.4.6 The Uncertainty-Aware Generative Model

Keith and O'Connor (2018) propose a Generative Probabilistic Modelling (GPM) approach to prevalence estimation. The proposed method directly conducts inference for the unknown prevalence and caters for confidence intervals (CIs) inference. CIs aim to capture the uncertainty of the model in providing an accurate class prevalence prediction (i.e., the more confident the model is about its prevalence estimation, the narrower the CI, and vice versa). This is the first quantification method in the literature that directly models uncertainty in terms of CIs.[15] CI are later discussed in more detail in Section 5.8.

The idea explored in Keith and O'Connor (2018) is to learn a generative probabilistic model that, by assuming (i) the documents be conditionally dependent on the label (i.e., the data generation process is of the form $\mathcal{Y} \rightarrow \mathcal{X}$, see Section 1.5), and (ii) that the class-conditional (unigram) language models remain invariant between training and unlabelled distributions, proceeds by first sampling a prior class distribution Θ, then sampling a label $y_i \sim$ Bernoulli(Θ) for each document,

---

[15] Note that, although CIs were already mentioned in the work of Hopkins and King (2010), their method (README – see 4.4.1) is not properly probabilistic, and CIs were obtained via bootstrap.

and finally sampling a bag-of-words document $x_i \sim$ Multinomial(ϕ_{y_i}) conditioned on the label. Different methods are explored as alternatives for the language model determining ϕ_{y_i}. In particular, two *explicit* ones (Multinomial Naive Bayes and Loglin) that directly model $p(\mathbf{x}|y)$, and another *implicit* (LR-Implicit) that instead estimates said class-conditional $p(\mathbf{x}|y)$ via the posteriors $p(y|\mathbf{x})$ generated by a discriminative classifier (a logistic regressor). The optimal prevalence (along with the CI) for a set of unlabelled items is then sought by simply exploring a grid of possible values and returning the one maximising the marginal log probability of all unlabelled documents.

Among the variants explored, LR-Implicit yields the best results in terms of MAE (for natural and artificial training prevalence values) and CI coverage (the proportion of times the $\text{CI}_{\alpha=0.1}$ happens to contain the true class prevalence).

Technically, a generative model equipped with a discriminative classifier as a proxy for computing $p(\mathbf{x}|y)$ from the posteriors $p(y|\mathbf{x})$ might better fit within the family of aggregative methods (discussed in Section 4.2) (indeed, the authors discuss the close connections between LR-Implicit and the aggregative SLD method of Saerens et al. (2002), see Section 4.2.9). However, the fact that the general generative framework described in Keith and O'Connor (2018) does only require the specification of a language model conditioned on the class labels (as directly attained by the *explicit* variants), squarely places the approach within the non-aggregative methods.

4.4.7 Deep Quantification Network

Qi et al. (2020) propose a Deep Quantification Network (DQN) that makes quantification predictions by combining quantification predictions made on samples from the test set to be quantified. More specifically, the training set L of labelled objects (binary or multi-class labels), is split, by sampling without replacement, in $\lfloor \frac{|L|}{m} \rfloor$ *m-tuples* of m objects (e.g., 100 objects). The training examples for DQN are thus the m-tuples with their prevalence values, as determined by the labels assigned to the elements in the tuple. The sampling policy that generates the m-tuples is a parameter of the method. The authors tested two sampling methods: random sampling, which may have low variance in the prevalence values of the m-tuples, thus risking to overfit DQN on the prevalence of the training set. The other sampling method is based on the Zipf distribution, which produces samples that exhibit more varied prevalence values, aimed at contrasting overfitting to the prevalence of the training set.

A set of m-tuples generated using the whole training set defines an *epoch* of the training process. Many set of m-tuples, and thus many training epochs, are used to train the DQN.

The DQN is composed of three main components, chained one after the other:

- A sample feature extraction component. This component produces a vectorial representation for each object in each of the m-tuples. This component is media-dependent; for text it can be an LSTM-based network, for images a CNN-based one.
- A m-tuple feature extraction component. This component converts the m sample feature vectors for an m-tuple in a single vector representing the whole m-tuple. The authors tested simple methods such as concatenation (CON), averaging (AVG), minimum (MIN), and maximum (MAX) of the sample vectors, as well as a dense layer with sigmoid activation function (NN).
- A class distribution prediction component. This component is implemented as a dense layer with output of the size of the number of labels, and softmax activation function, to output a probability distribution.

DQN is thus a feature extraction network followed by one (if CON, AVG, MIN, MAX components are used) or two (if NN is used) dense layers that convert the feature vector into prevalence estimations.

At test time the test size is split in m-tuples in the same way as the training data, prevalence predictions are collected for every m-tuple and averaged. Similarly to the training process, more than one split can be generated for a test set. In this case the various prevalence predictions from the different splits are averaged to produce the final prediction for the test set.

Qi et al. (2020) tested their method on binary (IMDb) and multi-class text datasets up to 20 labels (20 Newsgroups (Lang, 1995)), comparing it against CC, PCC, ACC, PACC, and ReadMe (King and Lu, 2008). The configuration using Zipf-based sampling and NN as the m-tuple feature extraction component always performed better than any other configuration, and better by a 45% on average, measured in terms of MAE reduction, than any of the compared baselines.

A key difference between DQN and QuaNet (Section 4.2.12) is that DQN directly tackles the quantification problem without leveraging on a classification method. In QuaNet, the document embeddings and classification scores are based on solving a classification problem. In DQN, all vector representations, including the feature vector representation for a single item in an m-tuple, are learned during the end-to-end learning[16] process that aims at quantification.

---

[16] In deep learning, the expression "end-to-end learning" indicates that all the parameters of a possibly complex and deep network are fitted at the same time during a single training phase, considering the whole network as a single model. This is in contrast to other training approaches in which some neural models are regarded as pre-trained models, and that typically consist of either training (only) a set of additional layers, or modules, stacked on top of the pre-trained model, or performing fine-tuning of the pre-trained model using the dataset at hand.

It is worth noting that setting the m-tuple size to the extreme, and rather odd, value of one transforms DQN into an aggregative method based on classification. When m-tuple size is one, any m-tuple can only have a prevalence of either one or zero, i.e., coinciding with the classification label of the single item it contains. The DQN thus classifies every single items and then outputs its prevalence estimate by looking at the set of classification scores. For any other m-tuple size value larger than one DQN can be considered a non-aggregative method.

Chapter 5
Advanced Topics

5.1 Ordinal Quantification

A special case of single-label classification is the ordinal one, in which the $m > 2$ classes are arranged in a total order. In this case, classes define a discrete, typically non-metric, qualitative scale. An example of this is the star rating model of product reviews, which is a typical problem faced in sentiment analysis. The sentiment scenario is one that highlights how quantification fits well with ordinal problems, as the typical use of ordinal ratings is to observe how the aggregated evaluations distribute among the various grades.

It is straightforward to observe that any quantification method for the SLQ case (see Section 4) can be applied to the ordinal case, and also that this approach is likely suboptimal as it does not take advantage of the total order among classes. Esuli and Sebastiani (2010b) discussed the scenario of ordinal quantification, and proposed an evaluation measure for it (see Section 3.2). The 2016 SemEval challenge proposed an ordinal quantification task (Nakov et al., 2016) that collected ten submissions from participants. Among them, only two submissions were based on methods specifically designed for the ordinal quantification task.

The method proposed by Da San Martino et al. (2016a,b), winner of the challenge, builds a binary tree from a set of binary classifiers trained on $(m - 1)$ split points of the ordinal scale. For example, when $m = 5$, four binary classifiers are trained: one that classifies elements in $\{y_1\}$ from the elements in $\{y_2, y_3, y_4, y_5\}$, and three other for the $\{y_1, y_2\}$ vs $\{y_3, y_4, y_5\}$, $\{y_1, y_2, y_3\}$ vs $\{y_4, y_5\}$, and $\{y_1, y_2, y_3, y_4\}$ vs $\{y_5\}$ splits. All the binary classifiers are corrected for quantification by applying PCC (see Section 4.2.2). The root node of the tree structure is determined by selecting the binary classifiers that has the smallest quantification error, measured via KLD. Subsequent nodes of the tree are determined recursively on the subsets of classifiers selected by the split of the parent node, until a split selects a single classifier. Quantification is performed by accumulating posterior probabilities for each element in the set of unlabelled items with respect

© The Author(s) 2023
A. Esuli et al., *Learning to Quantify*, The Information Retrieval Series 47,
https://doi.org/10.1007/978-3-031-20467-8_5

to each category. The posterior probability for an element with respect to a category is defined by the product of the probabilities in the path of the binary tree the goes from the root to the leaf associated with that category.

Esuli (2016) proposed a similar approach, in which a binary tree of classifiers is built on split points of the ordinal scale. The difference with the previous approach lies in the criterion used to define the tree, which in this work is based on selecting for the root (and then recursively for any other subtree) the split point that produces the most balanced training set, adopting the heuristic that quantification method may perform better on balanced dataset rather than unbalanced ones. For example, for on a ordinal scale that has labels $\{y_1, y_2, y_3, y_4\}$, respectively with 40, 20, 10 and 10 training examples, the best split for the root of the tree is $\{y_1\}$ vs $\{y_2, y_3, y_4\}$, as it produces a 50-50% split of the examples. The method of Da San Martino et al. (2016a), which is based on the actual evaluation of the quantification accuracy to define the binary tree, experimentally proved superior to the one of Esuli (2016).

5.2   Regression Quantification

Aggregative approaches can provide useful results also in applications where regression (not classification) is the task at hand for single data points. In a foundational work with little follow-up (Bella et al., 2014), the problem of quantification for regression is outlined, aimed at estimating composite quantities such as sales, quantities of consumed goods, or overall duration.

The authors provide a supporting sample application: "Consider a maternity ward that has collected data about baby weight at birth (dependent variable) for risk pregnancies, jointly with several features about the mother and her current and previous pregnancies (input variables). With these (training) data, a regression model has been trained in order to predict baby weight. In order to better plan the resources needed and the number of expected complications, the hospital wants to estimate the distribution of weight births for the following month, according to a new group of pregnant women (unlabelled data) that the maternity ward is monitoring for future deliveries."

Let y denote the dependent variable, as customary in regression settings. As a key aggregated value to quantify is the average of the dependent variable over a sample U of unlabelled items is considered. A first trivial solution is proposed by computing

$$\hat{\mu}_U = \mu_L \tag{5.1}$$

i.e., the regression counterpart of Maximum Likelihood Prevalence Estimation (Section 4.1), dubbed *Test to Train* (TT). As usual, L represents the labelled (training) set, and U the unlabelled (test) set.

Another solution which neglects dataset shift (Section 1.5) and performs simple aggregation of individual estimates, is dubbed *Regress and Sum* (RSu), and corresponds to computing

$$\hat{\mu}_U = \frac{\sum_{i=1}^{|U|} \hat{y}_i}{|U|} \tag{5.2}$$

where $\hat{y}$ represents the estimate provided by a regression model trained on L. This estimate is clearly reminiscent of Classify and Count. RSu estimates are likely to suffer from potential weakness of the underlying regression model, typically trained via minimisation of mean square error on L. The authors argue that quadratic loss functions discourage predictions far from the mean, thus bringing about more packed predictions.

This may be acceptable if we are only interested in a single value or indicator such as the mean $\hat{\mu}_U$, but becomes more of an issue if we are interested in estimating a full probability distribution for the output value y, a different and fully legitimate task in the realm of quantification for regression. To exemplify, the counterpart of RSu for this task can be computed as

$$\hat{P}_U(y \le r) = \frac{\sum_{i=1}^{|U|} \mathbb{1}(\hat{y}_i \le r)}{|U|} \tag{5.3}$$

where $\mathbb{1}(\cdot)$ is the indicator function. This method is dubbed *Regress and Splice* (RSp).

A further drawback of RSu is the inheritance of bias from its underlying regression model, which can be non-zero even in the absence of dataset shift. The authors propose three heuristics designed to reduce the impact of the above-mentioned issues thus improving aggregate quantification:

- *Adjustment* is aimed at compensating for bias, as estimated on L. This leads to a method dubbed *Adjusted Regress and Sum* (ARS), summarised by the formula

$$\hat{\mu}_U^{ARS} = \hat{\mu}_U^{RSu} + \alpha B_L^{RSu} \tag{5.4}$$

Here B_L^{RSu} is the bias of the RSu estimate computed on L, and α represents a modulating factor optimised empirically.

- *Segmentation* responds to the need for different adjustments across regions of the input space. In other words, it is reasonable to expect that the bias of a regression model will be region-dependent, bringing about systematic underestimation in some areas, while overestimating elsewhere. A number of thresholds are suitably defined for y, based on values taken by y in the training set L. Predictions $\hat{y}$ issued on U are binned according to these thresholds, approximated with a value deemed representative of the respective bin, and adjusted in a bin-dependent way. More in detail, the computation is the result of the following steps:

1. Thresholds are selected based on three alternative criteria, namely equal width of intervals, equal frequency (i.e., in such a way that the resulting partition on L determines sets of same cardinality), or k-means.
2. After partitioning L based on variable y, the values of the respective estimates $\hat{y}$ from a single bin are averaged, in order to determine a prototypical value $\hat{y}^m$ for said bin.
3. After performing regression on U, each data point is assigned to a bin via comparison of $\hat{y}$ with bin thresholds. Each regression estimate is then replaced by its prototype $\hat{y}^m$.
4. Finally, adjustment is performed independently on each bin.

Individual predictions are thus corrected according to formula

$$\hat{y} = \hat{y}^m_j + \alpha B_j \qquad (5.5)$$

where bin membership is denoted by subscript j. Finally, $\hat{\mu}_U$ is computed as the average of predictions over U.

- *Spreading* is aimed at counteracting the compression of predictions $\hat{y}$, brought about by regression models which have a tendency to produce packed outputs. For this reason, estimates $\hat{y}$ are corrected via the Nadaraya-Watson kernel as a first step. This kernel smoothing algorithm allows to artificially increase the variance of predicted values to better match the variance of the real values y when required. Spreading can be used in conjunction with all techniques described above, including TT, RSu and ARS. It is deemed especially useful when the task at hand requires an estimate of the whole probability density, less so when the interest lies in the average value $\hat{\mu}_U$.

5.3 Cross-Lingual Quantification

Cross-Lingual Quantification (CLQ) is the task of performing quantification in scenarios in which training documents in the target language for which quantification needs to be performed do not exist (or are too few as to deploy a reliable quantifier) but exist for a different source language. Additionally, large quantities of unlabelled documents are assumed to be easily accessible for both domains. Esuli et al. (2020) formally defined the task and proposed preliminary baselines for binary sentiment classification. The key observation is that, when performed via aggregative methods, cross-lingual quantification could be directly enabled via the combination of cross-lingual classification and quantification correction. In Esuli et al. (2020), *Cross-lingual Structural Correspondence Learning* (Prettenhofer and Stein, 2011) and *Distributional Correspondence Indexing* (Moreo et al., 2016), two methods capable of generating cross-lingual vectorial representations (i.e., in a language-agnostic vector space), were used to train (general purpose) classifiers and tested in combination with CC, PCC, ACC, PACC, and QuaNet (discussed in Section 4.2).

Note that CLQ is an instance of *transfer learning* (Pan and Yang, 2010), the general learning framework dealing with differences in data distribution and data representation between the source and the target domains. Other variants of transfer learning (e.g., cross-domain text quantification) remain, to the best of our knowledge, unexplored. We are likewise unaware of more general CLQ methods tackling quantification by topic (instead of by sentiment), dealing with multi-class problems (instead of binary), or adopting non-aggregative approaches (that is, without relying on cross-lingual classification as an intermediate step).

5.4 Quantification for Networked Data

Networked data quantification is a special quantification setting where a network structure connects the individual unlabelled items, as is the case e.g., with hyper-linked web pages. In classification, the presence of hyperlinks allows the use of supervised ("relational") learning techniques that leverage both endogenous features (e.g., textual content) and exogenous features (e.g., hyperlinks and/or the labels of neighbouring items) (Chakrabarti et al., 1998; Macskassy and Provost, 2007). The term "collective classification" (see also Section 6.4) is often used to denote the fact that the classification of networked items is best tackled collectively, and not for each item in isolation of the others, since the label to be assigned to one item may influence the label to be assigned to another item. This is consistent with homophily effects and preferential attachment often seen in networked data. So, one obvious method of performing relational quantification is using a state-of-the-art collective classification algorithm and correcting the resulting prevalence estimates via method ACC (or Method Max, Method X, T50, MS, MM). Tang et al. (2010) follow this route by using the wvRN algorithm of Macskassy and Provost (2003) as the collective classification algorithm. However, they further propose a non-aggregative method called *Link-Based Quantification* (LBQ), inspired by the ACC method of Section 4.2.3. Let $p(\vec{i}^k)$ denote the fraction of nodes in the network that link to node i with $(k-1)$ levels of indirection (so that, e.g., $p(\vec{i}^1)$ is the fraction of nodes in the network that directly link to node i). From the law of total probability it follows that

$$p(\vec{i}^k) = p(\vec{i}^k|\oplus) \cdot p_U(\oplus) + p(\vec{i}^k|\ominus) \cdot (1 - p_U(\oplus)) \tag{5.6}$$

entailing

$$p_U(\oplus) = \frac{p(\vec{i}^k) - p(\vec{i}^k|\ominus)}{p(\vec{i}^k|\oplus) - p(\vec{i}^k|\ominus)} \tag{5.7}$$

Equation 5.7 allows estimating $p_U(\oplus)$, since the value of $p(\vec{i}^k)$ can be observed directly in the network, while the values of $p(\vec{i}^k|\oplus)$ and $p(\vec{i}^k|\ominus)$ can be estimated from a training set. A different estimate $\hat{p}_U^{(i,k)}(\oplus)$ of $p_U(\oplus)$ can be obtained for each pair (i, k) composed of a node i in the network and an integer value of k. In order to obtain a robust estimate, the authors compute all estimates for $k \in [1, k_{max}]$ (for a given k_{max}), and use their median as the final estimate $\hat{p}_U(\oplus)$. Quantification based on homophily is further explored in Milli et al. (2015). A community detection algorithm is run on the whole network graph (comprising elements from U and L). Each node in U is subsequently assigned the most frequent label from nodes in its community belonging to L. In case of community overlap, a prevailing one is identified based on its density or on highest class prevalence within the community. Alternatively, ego-networks are proposed as a way to define the community of a given node. Given a node's neighbourhood (nodes directly or k-hop-connected to it), its missing label is determined as the majority one in the neighbourhood.

After label assignment is carried out, Classify and Count and Adjusted Classify and Count are employed as strategies to aggregate the results. For the latter, false positive rates and true positive rates are estimated on L with a leave-one-out approach.

5.5   Cost Quantification

A specific flavour of quantification has been tackled by Forman (2006, 2008) and dubbed *cost quantification*. For this application, each data point comes with additional cost information associated to it. A key application is represented by a business looking for insight into warranty costs for its products. Given a set of customer support logs, comprising textual data about issues described by customers and the cost of support (e.g., repairs), we are interested in quantifying how much each type of issue is contributing to after-sales expenses. Classes are represented by different issues or any atomic feature that might drive quality assurance decisions for the business, e.g., **CrackedScreen** or **SwollenBattery**. This task is trivially resolved by a quantifier if the average cost for a given issue is fixed and known in advance. However, a further source of complexity is often introduced due to variability of prices for components.

Classify and Total (CT), is the simplest algorithm considered. Being the counterpart of Classify and Count, it is based on running a classifier on each sample from U and adding up the cost $c(\mathbf{x})$ associated to each sample labelled as belonging to the class of interest, which comes down to computing

$$S_y = \sum_{\mathbf{x} \in U : h(\mathbf{x}) = y} c(\mathbf{x}) \tag{5.8}$$

This approach has similar limitations to Classify and Count.

Grossed-Up Total (GUT) mitigates this problem by pushing the CT estimate S_y upwards or downwards according to the ratio between the class prevalence estimate by a proper quantifier M_q and the one provided by the classifier employed, i.e.,

$$S'_y = S_y \times \frac{\hat{p}_U^{M_q}(y)}{\frac{1}{|U|}\sum_{x \in U} \mathbb{1}(h(\mathbf{x}) = y)} \tag{5.9}$$

which can be rewritten as

$$S'_y = \hat{p}_U^{M_q}(y)|U| \times \frac{S_y}{\sum_{x \in U} \mathbb{1}(h(\mathbf{x}) = y)} \tag{5.10}$$

thus making two factors explicit. The first represents an estimate of cardinality for class y within U given by quantifier M_q, while the second one can be interpreted as a best guess of average cost for class y provided by classifier $h(\mathbf{x})$, which, however, is quite likely to be polluted by misclassified items.

*Conservative Average * Quantifier* (CAQ) is aimed at reducing pollution by computing a cost average on a predefined amount of items from U, which we deem very likely to belong to class y. These items are taken in decreasing order of posterior probability $p(y|\mathbf{x})$.

*Precision Corrected Average * Quantifier* (PCAQ) takes the above idea a step further by estimating the precision (or Positive Predictive Value – PPV) of classifier $h(\mathbf{x})$ on the unlabelled set U. For ease of notation, in the binary case, let us shorten the symbol for estimates of prevalence for class $\oplus$ within U provided by quantifier M_q to $q = \hat{p}_U^{M_q}(\oplus)$. Moreover, let PPV_h denote the precision of classifier $h(\mathbf{x})$ on U. The values of PPV_h on U can be computed from estimates of class prevalence q and estimates of true and false positive rates for $h(\mathbf{x})$ (TPR_h, FPR_h), obtained via cross-validation on L, i.e.,

$$\text{PPV}_h = \frac{q \cdot \text{TPR}_h}{q \cdot \text{TPR}_h + (1-q) \cdot \text{FPR}_h} \tag{5.11}$$

This value is then employed to compute the average cost of positive predicted instances via

$$C_\oplus^h = \text{PPV}_h C_\oplus + (1 - \text{PPV}_h)C_\ominus \tag{5.12}$$

where $C_\oplus$ is the average cost of items in class $\oplus$, which we need to estimate. A further equation linking these quantities can be specified on the whole set U of unlabelled items, i.e.,

$$C_U = p_U(\oplus)C_\oplus + (1 - p_U(\oplus))C_\ominus \tag{5.13}$$

where C_U is the average cost of items in U. After solving for $C_\ominus$, plugging into Equation 5.12, and substituting $p_U(\oplus)$ with its estimate q, we obtain

$$C_\oplus = \frac{(1-q)C_\oplus^h - (1-\text{PPV}_h)C_U}{\text{PPV}_h - q} \tag{5.14}$$

which is then multiplied by estimated class cardinality $q \cdot |U|$ to get the final cost quantification. Note that both estimates of classifier precision PPV_h and average cost $C_\oplus^h$ depend on how the classifier's threshold is selected.

Median Sweep of PCAQ applies the philosophy of Median Sweep from Section 4.2.6 to PCAQ by considering several values for classifier threshold, getting a different estimate C_y for each of them via PCAQ, and regarding their median as a final estimate.

*Mixture Model Average * Quantifier* applies a similar idea directly to Equation 5.12. By letting threshold t vary we obtain

$$\frac{C_\oplus^t}{\text{PPV}^t} = C_\oplus + C_\ominus \frac{1 - \text{PPV}^t}{\text{PPV}^t} \tag{5.15}$$

i.e., a system of equations, one for each threshold value, which can be solved for $C_\oplus$, $C_\ominus$ via linear regression.

Note that these methods approximate the values of TPR and FPR on the unlabelled set U with estimates computed via cross-validation on L, which may be a bad approximation unless $p_L(\mathbf{x}|y) = p_U(\mathbf{x}|y)$, i.e., unless L and U are connected by prior probability shift.

5.6   Quantification in Data Streams

Yang and Zhou (2008) consider the problem of estimating the shift in prior distribution while observing a sequence of objects from a stream. Their aim is to improve the classification accuracy by using shift updated priors in the classification model that is trained only once at the beginning of the process, i.e., without resorting to active learning and retraining. The proposed method adapts the EM method of Saerens et al. (2002) to work from a batch setup, i.e., estimating new priors for a set of unlabelled objects, to an online setup, i.e., correcting priors every time a new object appears in the stream. Differently from the method by Saerens et al. (2002), the Online EM (OEM) method of Yang and Zhou (2008) applies the E and M steps only once to each element that is sequentially generated by the stream. The initial priors, as well as the likelihood function, are computed on a training set. The E step computes the posteriors probabilities of the k-th element of the sequence $\mathbf{x}_1 \ldots \mathbf{x}_n$ of elements of the set U of unlabelled items using the likelihood function and the priors for the k-th step, similarly to the method by Saerens et al. (2002). The M step computes the corrected priors for the next $k+1$ element of the sequence using an exponential forget function that combines the priors of the k-th step with the posteriors of the k-th element, i.e.,

$$\hat{p}_{k+1}(y) = \alpha\,\hat{p}_k(y|\mathbf{x}_k) + (1 - \alpha)\hat{p}_k(y) \tag{5.16}$$

The OEM method is thus an online quantification method in the strict sense of online processing, as each element of the sequence is observed and processed only once.

In experiments OEM performs better than the original EM at improving the classification accuracy, yet the actual priors' estimation are not very accurate. Zhang and Zhou (2010) observed that this issue is likely related to a small-sample effect, i.e., that priors update in Equation 5.16 is determined by a single element. They propose to overcome this issue by means of a transfer estimation method, which computes the M step using the posteriors from N previous elements in the stream, i.e, Equation 5.16 is changed into

$$\hat{p}_{k+1}(y) = \alpha\frac{1}{N}\sum_{i=0}^{N-1}\hat{p}_k(y|\mathbf{x}_{k-i}) + (1 - \alpha)\hat{p}_k(y) \tag{5.17}$$

Maletzke et al. (2018) explore the use of active learning on data streams as a device to improve the quantification accuracy. They define data streams as generators of instances across time. For quantification, they consider U to be composed of a sequence of event windows U_t across time. Quantification requests happen whenever an event window is complete. The true label y is known for an initial batch of instances that define the training set L. The true label for successive instances may be available after a verification latency time T_l, which may range from $T_l = 0$ to $T_l = \infty$. The first case means that, if requested, the true label for an instance is immediately available. This is an unrealistic case for most real-world applications as some time is inevitably required by the labelling oracle, typically a human annotator, to produce the true labels. The latter case of $T_l = \infty$ means that no true labels will be ever available for instances outside the training set, which is an extreme scenario in which no active learning strategy can be applied. Active learning can be exploited in all the cases for which $T_l < \infty$, exploring many possible strategies and trade-offs between labelling cost and quantification accuracy improvement.

The methods proposed by Maletzke et al. (2017, 2018) are template methods as they leverage a classification-based method to perform the actual quantification, while they manipulate the training data (transforming or enriching it).

The *Stream Quantification by Score Inspection* (SQSI) algorithm (Maletzke et al., 2017) leverages statistical tests to decide if a classifier trained on L can be reliably used to perform classification and quantification on U_t. The algorithm works as follows:

1. It starts by training a classifier h on an initial training set L.
2. Given a set of items U_t to quantify, h is used to get the classification scores on all of them.

3. The set of classification scores on U_t is compared to the set of classification scores on L (obtained with a leave-one-out cross validation). The comparison is done with a Kolmogorov-Smirnov test, under the null hypothesis that the two sets of scores come from the same distribution.

   (a) If the null hypothesis is not rejected, a quantification method based on h is used to estimate class prevalence on U_t. The algorithm repeats from Step 2 for the successive set U_{t+1}.
   (b) Otherwise, the algorithm makes a first attempt at transforming L into a shift adapted training set L' using the shift adaptation algorithm described in dos Reis et al. (2016).

4. h is replaced with a new classifier trained on L'.
5. The Kolmogorov-Smirnov test between L' and U_t classification scores from Step 3 is repeated.

   (a) If the null hypothesis is not rejected, a quantification method based on h' is used to estimate class prevalence on U_t. L' replaces L and the algorithm repeats from Step 2 for the successive set U_{t+1}.
   (b) If the null hypothesis is rejected again then the true labels of U_t are asked to an oracle, defining a new training set L. The algorithm repeats from Step 1 for the successive set U_{t+1}

Assuming a small shift between successive sets of items U_t, U_{t+1} one can expect that the oracle will seldom be consulted. In the experimental evaluation of Maletzke et al. (2017), performed on fourteen datasets with a very low number of features (only two features for 8 synthetic datasets, and less than 100 in the other cases), the portion of items labelled by the oracle was below 10% in all but one case.

The SQSI algorithm can help the quantification process only when the observed shift is within the range of correction of the shift adaptation method, otherwise it fails, requiring a complete labelling of the set of items to be quantified by the oracle. The SQSI-IS (where IS stands for Instance Selection) algorithm tries to reduce the amount of labelling required by using instance selection and self-learning whenever the shift adaptation method fails. Instead of requiring the oracle to label the whole set U (Step 5b above), only a fraction of elements of U is selected for labelling by the oracle, while the remaining part is labelled using an iterative process of self-learning adding to L the element of $U \setminus L$ that is classified with the highest confidence. The authors test several instance selection methods (random, clustering based, farthest-first traversal), and find that a clustering-based approach performs consistently better, with the best overall quantification performance observed for SQSI-IS instantiated with clustering and the PCC quantification method.[1] The observed reduction in labelling requests from SQSI to SQSI-IS is 50% on average, while achieving the same quantification performance.

---

[1] Maletzke et al. (2018) tested CC, PCC and ACC as the base quantification methods.

5.7 One-Class Quantification

A one-class classification problem assumes that the labelled examples are all positive examples of a single class, and that no negative examples are available. Performing quantification in the one-class case is challenging because it is not possible to measure a real prevalence on the training set L. Moreover, for quantification methods that rely on classification, also the one-class classification scenario is obviously a harder problem than the traditional classification scenario in which one has representative examples of both the positive and the negative classes.

Nonetheless, approaching a quantification problem as a one-class quantification problem may be a more robust approach in cases in which the definition of the negative cases is open. In a one-class setup the positive label will likely identify a specific property while the negative label comprises the universe of data points for which such property does not hold. In this case is it thus hard to have the domain of negative examples properly represented in the training set. The domain of negative examples may change considerably after training the quantification model. For example, one may be interested in training a **Sports** news quantifier, having as negative example only news about **Health**. The trained quantifier may be then applied to datasets that include news about **Economics** and **Politics**. In this scenario, a one-class quantifier, trained only on positive examples for **Sports**, may be more robust to the variation of data composition between the training phase and the deployment phase.

Moreira dos Reis et al. (2018a) propose two methods for one-class quantification, the *Passive Aggressive Threshold ACC* (PAT-ACC) the *One Distribution Inside* (ODIn) method, which draws inspiration from the MM approach (Forman, 2008, see Section 4.2.8). Both methods are designed to work in combination with one-class classifiers.

PAT-ACC extends ACC to work on one-class problems by observing that the problem of estimating FPR can be circumvented by choosing a conservative classification threshold, so that one can assume that FPR ≈ 0. If the classification threshold is set so that a quantile q of observations is classified as positive, then the TPR can be estimated as TPR $= 1 - q$, allowing to perform quantification using the ACC method (see Equation 4.5), i.e.,

$$\hat{p}_U^{\text{PAT-ACC}}(\oplus) = \min\left(1, \frac{p_U(h(\oplus))}{(1-q)}\right) \tag{5.18}$$

Moreira dos Reis et al. (2018a) claim that the PAT-ACC method is not sensitive to the value of q and report that a value of $q = 0.25$ is a generally good choice. They also suggest that an approach similar to Median Sweep can be adopted to avoid using a fixed q value.

The ODIn method compares the score distribution that is available only for positive examples in the case of the training set L with the score distribution for U, which includes both negative and positive examples. Scores from the classification of L are used to define a variable-width histogram H^L in which each bin has the same number of elements. The number of bins b is a parameter, which in Moreira dos Reis et al. (2018a) is set to $b = 10$. Scores from the classification of U define a histogram H^U, which uses the bin definition of H^L. The overflow of H^L in H^U is defined as

$$\text{OF}(\alpha, H^U, H^L) = \sum_{i=1}^{b} \max(0, H_i^U - \alpha H_i^L) \tag{5.19}$$

The value α scales the histogram H^L and OF measures how much the scaled histogram still has higher valued bins than H^U. Intuitively ODIn searches for the largest parameter α that better fits H^L inside H^U, then producing the quantification estimate by correcting it for its overflow, i.e.,

$$\hat{p}_U^{\text{ODIn}}(\oplus) = s - \text{OF}(s, H^U, H^L) \tag{5.20}$$

where

$$s = \sup_{0 \leq \alpha \leq 1} \{\alpha | \text{OF}(\alpha, H^U, H^L) \leq \alpha \mathcal{L}\}$$

where $\mathcal{L}$ is a parameter of the method. In Moreira dos Reis et al. (2018a) the authors set $\mathcal{L} = \hat{\mu}_{\text{OF}} + d\hat{\sigma}_{\text{OF}}$, where the values $\hat{\mu}_{OF}$ and $\hat{\sigma}_{OF}$ are the mean and standard deviation of the OF function estimated on pairs of samples from L, and d is a new parameter that replaces $\mathcal{L}$. The authors claim that the parameter d has a clearer semantic than $\mathcal{L}$, i.e., d is the number of standard deviations of the expected average overflow, and arbitrarily set to $d = 3$ for all of their experiments.

The problem of class prior estimation in the one-class case is faced in du Plessis and Sugiyama (2014). This work has the main goal of learning a classifier from positive examples and unlabelled data, and quantification is not the subject of its proposal. Yet, the proposed method, which they call PE, performs the estimation of class priors, considering it a necessary step to learn a good classifier. Given that the correct estimation of class priors is indeed quantification, we consider this work relevant to our goals. They start from the input density formula

$$q(\mathbf{x}; \Theta) = \Theta p(\mathbf{x}|\Phi(\mathbf{x}) = \oplus) + (1 - \Theta)p(\mathbf{x}|\Phi(\mathbf{x}) = \ominus) \tag{5.21}$$

observing that $q(\mathbf{x}; \Theta) = p(\mathbf{x})$ when $\Theta = p(\oplus)$, thus defining a full-matching method for prior estimation. However, in the one-class case $p(\mathbf{x}|\Phi(\mathbf{x}) = \ominus)$ is unknown. To overcome this issue the authors make the assumption that the class-conditional densities $p(\mathbf{x}|\Phi(\mathbf{x}) = \oplus)$ and $p(\mathbf{x}|\Phi(\mathbf{x}) = \ominus)$ are not strongly overlapping and propose a partial-matching estimation method. Such method matches only $\Theta p(\mathbf{x}|\Phi(\mathbf{x}) = \oplus)$ to $p(\mathbf{x})$ using the Pearson Divergence (PD), i.e.,

$$\hat{p}_U^{\text{PE}}(\oplus) = \arg \min_{\Theta} \text{PD}(\Theta) \tag{5.22}$$

where PD is defined as

$$\text{PD}(\Theta) = \frac{1}{2} \int \left(\frac{\Theta p(\mathbf{x}|\Phi(\mathbf{x}) = \oplus)}{p(\mathbf{x})} - 1 \right)^2 p(\mathbf{x})dx \tag{5.23}$$

The authors experimentally proved that the partial-matching method based on PD has a lower error than the method based on Equation 5.21 for the one-class case. In a subsequent work (du Plessis et al., 2017) the approach is further extended to other divergence functions.

Zeiberg et al. (2020) proposed the DistCurve algorithm that estimates the prevalence of a sample σ by leveraging of the concept of distance curve. A distance curve is computed starting from a sample σ and a labelled set L that contains only positive elements. Points of the curve are determined by sampling, with replacement, a random element from L, and measuring its distance from the closest element in σ, that element is removed from σ. The procedure continues until σ is empty. The idea is that the distance curve should show a steep increase in distance at the step $p_\sigma(\oplus)|\sigma|$, as all the positive elements have been removed from the set. A neural network is trained on distance curves generated on samples with known priors, so as to be able to predict the $\hat{p}_\sigma$ value from the distance curve for σ. In order to be robust to statistical variation caused by the sampling mechanism, the distance curve for σ that is given as input to the neural network is determined as the average of multiple runs of the method that computes the distance curve.

5.8   Confidence Intervals for Class Prevalence Estimates

A *confidence interval* (CI), in the context of quantification, is a range of values (l, h) which should contain the true prior probability $p_U(y)$ for class y with a desired level of confidence, such as 95%. In mathematical terms, l and h should be such that the probability of event $(p_U(y) \in (l, h))$ is equal to 0.95. This information is often more useful than a point estimate of class prevalence $\hat{p}_U(y)$.

Hopkins and King (2010) first mentioned computing bootstrapped CIs for their estimates, without providing much detail. CIs for quantification have received more attention in recent years. Keith and O'Connor (2018) propose a generative model, whose characteristics naturally allows for the computation of CIs for class prevalence values (Section 4.2.8). Let $p_U(\oplus)$ denote the true proportion of positives in U. Algorithms which support *Maximum a posteriori* estimation are typically used to compute the single most plausible value for $p_U(\oplus)$, i.e. the one that is most compatible with the covariates observed in U, but also support the computation of likelihood values for any possible $p_U(\oplus) \in [0, 1]$. The authors exploit this idea, training different versions of the generative models. At inference time, they employ grid search over all possible (quantised) values of $p_U(\oplus)$, in conjunction with a uniform prior, constructing a posterior density from which confidence intervals are derived.

Daughton and Paul (2019) propose a technique called *error-adjusted bootstrap* to compute CIs for quantification based on the outputs of a classifier, with a correction procedure accounting for its (im)precision. In the construction of a bootstrap sample, they draw an instance with covariates $\mathbf{x}$ from U, and feed it to a classifier $h(\mathbf{x})$, to obtain a predicted class $c \in \{\oplus, \ominus\}$. The bootstrap sample is expanded by using the classifier output as a parameter to sample from a Bernoulli distribution with success probability $p_U(\oplus|h(\mathbf{x}) = c)$; (un)successful draws result in attaching class $\oplus$ ($\ominus$) to the sample. Prevalence estimates for a single bootstrap sample are subsequently obtained by computing the frequency of $\oplus$ within it. Confidence intervals at a desired level are then constructed customarily, based on the estimates from all bootstrap samples. Crucially, the precision-related parameter $p_U(\oplus|h(\mathbf{x}) = c)$, shaping the Bernoulli distribution, is estimated on the training sample L. As duly noted by Tasche (2019), this approach does not generally work under dataset shift. This is due to the fact that $p_U(\oplus|h(\mathbf{x}) = c) = p_L(\oplus|h(\mathbf{x}) = c)$ is not guaranteed to hold. Hence, the approach of Daughton and Paul (2019) seems suited to handle covariate shift, a setting where the previous equation holds true.

Fernandes Vaz et al. (2019), whose work is discussed in Section 4.2.7, provide a central limit theorem for the ratio estimator, from which confidence intervals can be computed without any numerical simulation.

Tasche (2019) deploys a simulation study to shed some light on the topic of CIs in quantification tasks, under prior probability shift. Despite lacking the complexity of real-world datasets, the study provides some illustrative and interesting results in a controlled setting described very clearly. Several quantification methods are selected based on Fisher-consistency (Tasche, 2017) and popularity in the literature, including ACC (Section 4.2.3), PACC (Section 4.2.4), MS (Section 4.2.6), HDy (Section 4.2.8). Each of these methods is tested in a variety of settings, with probability shift ranging from strong to mild, exploiting underlying classifiers of variable discriminatory power, and testing on unlabelled samples of size $|U| \in \{50, 500\}$. For each combination of the above parameters, CIs at 90% are constructed via regular bootstrapping. One key finding is that, if a quantification method is based on an underlying classifier with high power, then the CIs will be shorter and more informative while retaining desired coverage levels.

The study also points out that, for quantification problems, prediction intervals are, in principle, more useful than confidence intervals. Indeed, a practitioner is not exactly interested in having a range for the true prior probability from which the unlabelled sample U originated, i.e., the target of confidence intervals. Rather, they plausibly care about having a range of plausible values for the *realised prevalence*, i.e. the percentage of points from U that belong to the positive class, a quantity that should be targeted by (more conservative) prediction intervals. However, the results of simulations carried out by Tasche (2019) in a variety of settings suggest that, for $|U| > 50$, as reasonable in most practical applications, the construction of confidence intervals is sufficient (adequate coverage) and there seems to be no need for the construction of more conservative prediction intervals.

Thanks to central limit theorems (see e.g., Section 4.2.7), confidence intervals for some approaches can be constructed without bootstrapping. Tasche (2019) also tests the effectiveness of this approach, concluding that it results in suboptimal results (e.g. low coverage) in the presence of certain conditions. As an example, if the true positive rate and false positive rate of an underlying classifier have to be estimated, a limited sample size for L may be a source of imprecision in said estimate, corrupting prevalence estimates and bringing about confidence intervals of insufficient size.

More recently, Denham et al. (2021) note that PCC can natively provide confidence intervals, since PCC may be thought of as computing the mean of a Poisson binomial distribution of the posterior probabilities (scaled by a constant factor), and since we know how to derive reliable confidence intervals under this assumption. The authors exploit this idea, along with other assumptions on the underlying distributions of a mixture model, to derive confidence intervals for their method GSLS (explained in Section 4.2.8).

Chapter 6
The Quantification Landscape

6.1 Historical Development

6.1.1 The Trajectory of Quantification

The "prehistory" of quantification research may be traced to the interest in the estimation of class prevalence from screening tests, as carried out in epidemiology. Accordingly, the first recorded "quantification" technique is probably the one of Gart and Buck (1966) (see Section 4.2.3). This literature is different from that discussed in the rest of this book (and this is the reason why the term "quantification" above is in quotation marks) since no training data (and no supervised learning) is involved here: the role of the classifier is here played by a clinical test that has imperfect (but known) sensitivity and specificity (see Section 6.4 for details). The estimation of class prevalence has remained an important concern of epidemiological research to this day, and several papers on this topic (e.g., Levy and Kass, 1970; Lew and Levy, 1989; Morvan et al., 2008; Rahme and Joseph, 1998; Viana et al., 1993; Zhou et al., 2002) have continued to appear in epidemiology-related journals to this day.

The first stage of such history in which supervised learning is involved coincides with interest in the estimation of class prevalence from the machine learning community, where the goal is (as already discussed in Section 2.1) that of building classifiers that are robust to the presence of distribution shift, and that are better attuned to the characteristics of the data to which they need to be applied. Here, the precursors seem to have been Vucetic and Obradovic (2001), but the most influential paper to date in this field is certainly that of Saerens et al. (2002); later works are, e.g., Alaíz-Rodríguez et al. (2011), Chan and Ng (2005), Chan and Ng (2006), Xue and Weiss (2009), and Zhang and Zhou (2010). As mentioned in Section 2.1, in this stream of research the estimated class prevalence values are not interesting *per se*, but only serve the purpose of allowing a better estimation of the posterior

© The Author(s) 2023
A. Esuli et al., *Learning to Quantify*, The Information Retrieval Series 47,
https://doi.org/10.1007/978-3-031-20467-8_6

probabilities $p(y|\mathbf{x})$ (and hence a more accurate classification) for unlabelled data in contexts characterised by significant distribution shift.

The second and last stage of such history coincides with interest from data mining, text mining, and content analysis; it is mainly the applications from these fields that have provided the impetus behind the most recent wave of research in quantification.

Some 10 years before the very first developments in this line of research, Lewis (1995, §7) had already evoked a task (that he called *counting*) that was to consist of simply counting the unlabelled items that belonged to a given class (which, once the counts are normalised by the total number of unlabelled items, coincides with quantification). In that paper Lewis observed that "if our goal is to count class members, and if we have estimates of the probability of class membership, we should use the estimates directly to estimate the number of class members, rather than use them to classify documents"; this is exactly the principle that Bella et al. (2010), unaware of Lewis' observation 15 years earlier, based their PCC method upon (see Section 4.2.2). In this work, Lewis briefly discussed a potential evaluation measure for "counting" (which consisted of the square of the differences between FP and FN), but did not discuss the task in any further detail. His remarks about "counting" went essentially unnoticed, and quantification had to wait another 10 years in order for someone to call attention to the need to study it as a task separate from classification.

This finally happened with (Forman, 2005) and the papers by the same author that soon followed (Forman, 2006, 2008; Forman et al., 2006); it is in these papers that the term "quantification" was coined, a term that has since stuck and become standard terminology. Contrary to the works mentioned above (re: "first stage of such history"), in these works the estimated class prevalence values are the true objects of interest. These works eventually became well-known among, and inspired, researchers in machine learning, data mining, and text mining to develop the new methods and algorithms that we have discussed in Sections 4 and 5.

There is one chapter in the history of quantification research that has yet to be written, though, i.e., the one on a widespread uptake of quantification technology by users, that unfortunately has yet to happen. One only needs to look at the proceedings of, say, recent computational social science conferences, to realise how many works are carried out where classification is used despite the fact that the investigators are only interested in results at the aggregate level. Undoubtedly, this has to do with a scarce awareness, on the part of data scientists, that prevalence estimation is not just a by-product of classification. It is a goal of this book to improve this awareness.

6.1.2 Shared Tasks

To the best of our knowledge, the only shared task that has gathered researchers on a challenge that explicitly addressed quantification is the "Sentiment Analysis in Twitter" task of the SemEval-2016 (Nakov et al., 2016) and SemEval-2017 (Nakov et al., 2017) evaluation campaigns. The general goal of this task was to evaluate algorithms that classify tweets by sentiment. In both 2016 and 2017, this task included a binary quantification subtask (where positive vs. negative attitudes towards the designated object had to be identified) and an ordinal quantification subtask (where these attitudes had to be graded on an ordered scale of five values). That a shared task devoted to sentiment classification in Twitter should include subtasks devoted to quantification is just natural, given the fact that (as already mentioned in Section 2.3) most researchers and practitioners who apply sentiment classification technology to Twitter datasets are essentially interested in aggregate results.

One fairly disappointing result of those subtasks was that most participants used Classify and Count solutions, albeit often based on some sophisticated sentiment classification technology using deep learning. This testifies to the fact that, despite its many potential applications, quantification is still a fairly unknown task, and that there is very little awareness that Classify and Count delivers suboptimal quantification accuracy.

An ongoing challenge at the time of writing this book is the LeQua 2022 lab on Learning to Quantify (Esuli et al., 2022). The challenge brings *textual* quantification into focus, and comes with 2 separate tasks: "T1" for binary quantification, and "T2" for single-label quantification. Each of the tasks admits two variants, one in which documents come in the form of dense vectors, and another where documents come in raw form. The datasets consists of product reviews from the Amazon website. Task "T1" consists of predicting the binary class prevalence of the sentiment polarity of the reviews, while task "T2" consists instead of predicting the class prevalence of the merchandise categories ("Automotive", "Baby", "Beauty", ...) of the products, for a total of 28 categories. While the training samples reflect the natural prevalence as from the Amazon website, the validation and test samples are generated following artificial prevalence values, according to the Kraemer sampling algorithm discussed in Section 3.4. The results of the challenge will be presented at the CLEF 2022 conference.

6.2 Software

6.2.1 Publicly Available Implementations

Throughout the second phase of the history of quantification (Section 6.1), especially in recent years, several works have been published that make software implementations public, thus favouring the reproducibility and, more broadly, the adoption of quantification techniques. Indeed, publishing a software implementation of a method proposed in a paper produces many benefits to research, e.g., it provides a reference implementation, it allows peers to replicate the experiments, and it facilitates the comparison of the method with others in lab experiments. Some of the authors who have published papers on quantification methods have published software implementing their methods, and, sometimes, also of the methods they used as baselines. Table 6.1 reports on the available implementations of quantification methods, the papers where the link to the implementation is to be found, and the sections of this book in which the method is discussed.

6.2.2   QuaPy: A Comprehensive Framework for Quantification

The last of the packages in Table 6.1 is our own. It is called QuaPy (Moreo et al., 2021a), and was originally conceived as supplementary material accompanying this book. As such, it provides implementations of the main concepts discussed here, and using the same "jargon". Differently from other existing packages, QuaPy is not only a suite of methods, but an ecosystem for quantification, catering for model evaluation (including implementations for the most important evaluation measures), model selection (targeting quantification-oriented loss functions), and visualisation tools for analysing the experimental results (some examples are shown in Section 6.3.2). QuaPy also provides access to commonly used datasets, and implements a common interface to allow using other datasets. It is a Python-based open-source package with BDS-3-Clause licence that can be directly installed via pip.[1] It is extensible and in constant evolution, so that anyone can contribute new material via GitHub.[2]

Figure 6.1 shows a complete example of QuaPy's usage. In this example, the IMDb dataset of movie reviews is fetched (it is downloaded the first time) and vectorised using TFIDF weights. The example goes on by training a PACC quantifier that uses Logistic Regression as the probabilistic classifier. The quantifier hyperparameters (C and *class_weight* in this case, all coming from the classifier) are optimised via grid search using the artificial prevalence protocol for generating a maximum of 100 validation samples of 500 data items each (as indicated by *eval_budget* and by the environment variable *SAMPLE_SIZE*, respectively) out of a 25% held-out validation set and in terms of mean absolute error. The model is refitted on the entire training set once the hyperparameters have been optimised. Model training is then followed by model evaluation, by applying the artificial prevalence protocol anew, this time on the test set. The evaluation routine used in this example is one that generates a Pandas dataframe containing the error figures for absolute error, relative absolute error, and Kullback-Leibler divergence (see Figure 6.2).

---

[1] https://pypi.org/project/QuaPy/

[2] https://github.com/HLT-ISTI/QuaPy

Table 6.1 Software packages implementing quantification methods. **Boldface** indicates the main method proposed by the paper where the link to the software is to be found. The "Section" column indicates where the main method is discussed in this book. The lower block of the table lists software packages that are not directly linked to a specific method.

Methods	Language	URL	Paper	Section
RE, CC, EM	R	bit.ly/QuantSoft2	Fernandes Vaz et al. (2017, 2019)	4.2.7
HDx, **HDy**, CC, ACC, Max, X, T50, MS, MM, ACC, PCC, PACC, SLD	Java	bit.ly/QuantSoft3	Maletzke et al. (2017, 2019)	4.2.8, 5.6
QuaNet	Python	bit.ly/QuantSoft4	Moreo et al. (2021b)	4.2.12
SVM(NKLD), **SVM(KLD)**, SVM(Q)	C++	bit.ly/QuantSoft5a	Esuli and Sebastiani (2015)	4.3.1
ReadMe	R	bit.ly/QuantSoft1, bit.ly/QuantSoft6	King et al. (2013), Hopkins and King (2010)	2.6, 4.4.1
ReadMe2	R	bit.ly/QuantSoft7	Jerzak et al. (2022)	4.4.3
GPM	Python	bit.ly/QuantSoft8	Keith and O'Connor (2018)	4.4.6
ARC, CC, ACC, PCC, PACC	Python	bit.ly/QuantSoft9	Esuli (2016)	5.1
Cross-lingual QuaNet	Python	bit.ly/QuantSoft10	Esuli et al. (2020)	5.3
DistCurve	Python	bit.ly/QuantSoft11	Zeiberg et al. (2020)	5.7
PE	Matlab	bit.ly/QuantSoft12a	du Plessis and Sugiyama (2014)	5.7
ODIn, HDy	Lua	bit.ly/QuantSoft13	Moreira dos Reis et al. (2018a)	5.7
DQN	Python	bit.ly/QuantSoft16	Qi et al. (2020)	4.2.12
CC, ACC, Max, X, T50, MS, MM, ACC, PCC, PACC, SLD, PE	Matlab	bit.ly/QuantSoft14	N/A	N/A

(continued)

Table 6.1 (continued)

Methods	Language	URL	Paper	Section
CC, ACC, PCC, PACC, X, T50, Max, MS, DyS, ReadMe, HDx, HDy, SLD, CDE, SVM(KLD), SVM(Q), MM, QF, PE	Python	bit.ly/QuantSoft17	Schumacher et al. (2021)	N/A
CC, ACC, PCC, PACC, SLD, HDy, QuaNet, Ensembles, SVM(NKLD), SVM(KLD), SVM(Q), SVM(AE), SVM(RAE)	Python	bit.ly/QuantSoft15	Moreo et al. (2021a)	N/A

```
1    import quapy as qp
2    from quapy.method.aggregative import PACC
3    from sklearn.linear_model import LogisticRegression
4    import numpy as np
5    import pandas as pd
6
7    # setting this environment variable allows some
8    # error metrics (e.g., mrae) to be smoothed
9    qp.environ["SAMPLE_SIZE"] = 500
10
11   dataset = qp.datasets.fetch_reviews('imdb', tfidf=True, min_df=5)
12
13   # model selection with the APP
14   model = qp.model_selection.GridSearchQ(
15     model=PACC(LogisticRegression()),
16     param_grid={
17       'C': np.logspace(-4, 5, 10),
18       'class_weight': ['balanced', None]
19     },
20     protocol='app',
21     eval_budget=100,
22     error='mae',
23     refit=True, # retrain on the whole labelled set once done
24     val_split=0.25,
25   ).fit(dataset.training)
26
27   df = qp.evaluation.artificial_prevalence_report(
28     model, # the quantification method
29     dataset.test, # the set on which the method will be evaluated
30     n_prevpoints=101, # i.e., using the grid [0.,.01,.02,....,.99,1.]
31     n_jobs=-1, # the number of parallel workers (-1 for all CPUs)
32     random_seed=42, # allows replicating test samples across runs
33     error_metrics=['ae', 'rae', 'kld']) # evaluation metrics
34
35   print(f'best hyper-params={model.best_params_}')
36
37   pd.set_option('display.max_columns', None)
38   pd.set_option('display.width', 100)
39   print(df)
40
```

Fig. 6.1 Code example using QuaPy (version 0.1.6).

6.3 How Do Different Quantification Methods Fare?

6.3.1 A Tour of Experimental Results

In this section we show some of the most important quantification systems in action. This set of experiments is not intended to be exhaustive, nor is it intended to make conclusive statements about the relative merits of the different quantification systems being tested. The aim of this experimentation is rather that of demonstrating some of the major performance trends that typically arise naturally in different

```
best hyper-params={'C': 100.0, 'class_weight': 'balanced'}

    true-prev estim-prev ae rae kld
0   [0.0, 1.0]  [0.057592, 0.942407] 0.057592 28.824875 0.055245
1   [0.01, 0.99] [0.034542, 0.965457] 0.024542 1.127931 0.011950
2   [0.02, 0.98] [0.039174, 0.960825] 0.019175 0.466312 0.005742
3   [0.03, 0.97] [0.035338, 0.964661] 0.005339 0.088854 0.000428
4   [0.04, 0.96] [0.081784, 0.918215] 0.041784 0.531303 0.013911
..  ...  ...  ...  ...  ...
96  [0.96, 0.04] [0.948444, 0.051555] 0.011556 0.146937 0.001445
97  [0.97, 0.03] [0.972371, 0.027628] 0.002372 0.039477 0.000099
98  [0.98, 0.02] [0.967576, 0.032423] 0.012423 0.302125 0.002743
99  [0.99, 0.01] [0.967542, 0.032457] 0.022458 1.032138 0.010480
100 [1.0, 0.0]  [0.996870, 0.003129] 0.003129 1.566181 0.001716
```

Fig. 6.2 QuaPy's output example (version 0.1.6).

experimental settings. A more comprehensive overview and understanding of the relative merits of the different quantification systems might only be obtained by analysing the experimental evaluation carried out by multiple teams; see, e.g., Moreo and Sebastiani (2022), Pérez-Gállego et al. (2019), and Schumacher et al. (2021). The experiments we report here are extracted from Moreo et al. (2021a) and are obtained using the QuaPy framework.[3]

As the learning methods we chose CC (Section 4.2.1), PCC (Section 4.2.2), ACC (Section 4.2.3), PACC (Section 4.2.4), Forman's variants MAX (Section 4.2.5), MS and MS2 (Section 4.2.6), the mixture model HDy (Section 4.2.8), the expectation-maximisation-based SLD method (Section 4.2.9), SVM(AE) (Section 4.3.1) as the representative of the "explicit loss minimisation" family (minimising the same evaluation metric we use here), and E(HDy)$_{DS}$ as the representative of ensemble methods (Section 4.2.11); we set the number of base quantifiers to 30 and the number of members to be selected dynamically to 15 (we perform model selection independently for each base member).

The evaluation benchmark consists of 30 binary datasets coming from the UCI Machine Learning datasets, as were previously used by Pérez-Gállego et al. (2017). Results are mean AE scores (Section 3.1.3) obtained via 5-fold cross-validation. For each test fold, we follow an APP protocol (Section 3.4.2) and generate 100 different random samples of 100 instances each, using a grid of prevalence values $\{0.00, 0.05, \ldots, 0.95, 1.00\}$. The hyperparameters of the quantifiers are optimised via model selection for quantification (Section 3.5); in this case, minimising the

---

[3] The code to replicate all these experiments, and to generate the relative tables and plots, can be accessed via GitHub. See the files uci_experiments.py (runs all experiments), uci_tables.py (generates Table 6.2 directly in LaTeX), and uci_plots.py (generates plots from Figures 6.3, 6.4, 6.5, 6.6) included in the folder wiki_examples/ of the repository https://github.com/HLT-ISTI/QuaPy.wiki.git

Table 6.2 Values of AE obtained in our experiments; each value is the average across 10,500 test samples. **Boldface** indicates the best method for a given dataset. Superscripts † and ‡ denote the methods (if any) whose scores are *not* statistically significantly different from the best one according to a paired-sample, two-tailed t-test at different confidence levels: symbol † indicates $0.001 < p\text{-value} < 0.05$ while symbol ‡ indicates $0.05 \leq p\text{-value}$. For ease of readability, for each dataset we colour-code cells via intense green for the best result, intense red for the worst result, and an interpolated tone for the scores in-between.

| | Quantification methods | | | | | | | | | | |
	CC	ACC	PCC	PACC	MAX	MS	MS2	SLD	SVM(AE)	HDy	E(HDy)$_{DS}$
BALANCE.1	0.039	0.032	0.049	0.037	0.040	0.046	0.036	0.025	0.035	0.022	**0.020**
BALANCE.2	0.314	0.379	**0.264**	0.432	0.465	0.288	0.331	0.372	0.500	0.470	0.355
BALANCE.3	0.039	0.020	0.045	0.021	0.040	0.046	0.036	0.018	0.064	0.017	**0.014**
BREAST-CANCER	0.022	0.025	0.029	0.023	0.028	0.021	0.023	**0.020**	0.144	0.029	0.026
CMC.1	0.194	0.108	0.226	0.117	0.191	0.195	0.178	**0.094**	0.227	0.156	0.126
CMC.2	0.178	0.138	0.220	**0.098**	0.271	0.500	0.427	0.105	0.449	0.118	0.103
CMC.3	0.211	0.172	0.239	0.127	0.254	0.376	0.353	0.124‡	0.336	0.136	**0.122**
CTG.1	0.037	0.020	0.050	0.020	0.041	0.033	0.035	**0.017**	0.094	0.028	0.018
CTG.2	0.048	0.040	0.078	0.045	0.048	0.653	0.059	**0.030**	0.152	0.045	0.040
CTG.3	0.047	0.044	0.050	0.043	0.045	0.649	0.061	**0.022**	0.113	0.053	0.045
GERMAN	0.151	0.142	0.191	**0.092**	0.154	0.125	0.134	0.101	0.262	0.165	0.113
HABERMAN	0.231	0.190‡	0.237	0.267	0.242	0.572	0.244	**0.190**	0.283	0.399	0.324
IONOSPHERE	0.111	**0.074**	0.116	0.084	0.124	0.209	0.089	0.075‡	0.256	0.104	0.082
IRIS.2	0.201	0.241	0.195	0.183	0.251	0.412	0.256	0.215	0.461	0.075	**0.056**
IRIS.3	**0.019**	0.074	0.044	0.071	0.054	0.134	0.024	0.057	0.205	0.069	0.047

(continued)

Table 6.2 (continued)

	Quantification methods										
	CC	ACC	PCC	PACC	MAX	MS	MS2	SLD	SVM(AE)	HDy	E(HDy)$_{DS}$
MAMMOGRAPHIC	0.090	0.048	0.130	0.040	0.091	0.059	0.060	0.036	0.134	0.044	**0.031**
PAGEBLOCKS.5	0.048	**0.040**	0.067	0.041‡	0.066	0.474	0.115	0.070	0.342	0.085	0.066
SEMEION	0.042	0.049	0.058	0.040	0.038	0.500	0.074	**0.030**	0.070	0.037	0.047
SONAR	0.135	0.200	0.163	0.119	0.145	0.171	0.159	**0.114**	0.346	0.136	0.131
SPAMBASE	0.042	0.026	0.066	**0.022**	0.049	0.070	0.037	0.031	0.196	0.025	0.024
SPECTF	0.143	0.155	0.178	0.133	0.276	0.620	0.182	**0.105**	0.296	0.420	0.231
TICTACTOE	0.024	0.019	0.024	**0.014**	0.024	0.136	0.024	0.019	0.500	0.018	0.019
TRANSFUSION	0.178	0.139	0.215	0.097	0.220	0.510	0.433	**0.087**	0.442	0.246	0.166
WDBC	0.034	0.036	0.034	0.027	0.038	0.096	0.029	0.025	0.056	0.019	**0.015**
WINE.1	0.029	0.025	0.025	0.030	0.033	0.133	0.030	0.044	0.062	0.040	**0.019**
WINE.2	0.026	0.048	0.043	0.052	0.045	0.088	0.041	0.046	0.051	0.032	**0.022**
WINE.3	0.031	0.040	**0.016**	0.033	0.028	0.190	0.029	0.061	0.018†	0.018	0.025
WINE-Q-RED	0.140	0.076	0.183	0.059	0.141	0.065	0.099	**0.056**	0.222	0.065	0.058
WINE-Q-WHITE	0.150	0.077	0.194	0.064	0.149	0.113	0.124	**0.059**	0.247	0.072	0.066
YEAST	0.155	0.107	0.197	0.071	0.159	0.233	0.235	**0.066**	0.378	0.073	0.071
Average	0.104‡	0.093‡	0.121†	0.083‡	0.125†	0.257	0.132†	**0.077**	0.231	0.107†	0.083‡
Rank Average	5.733	4.967†	7.033	3.900‡	7.133	9.033	6.733	**3.133**	9.900	5.233	3.200‡

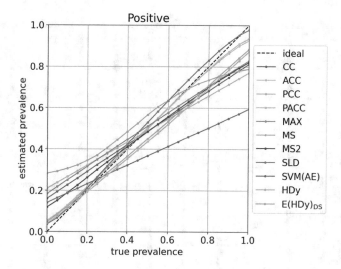

Fig. 6.3 Diagonal plot.

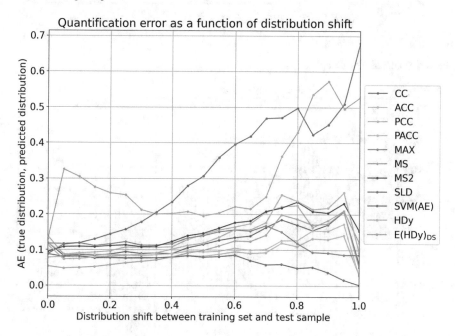

Fig. 6.4 Error-by-Shift plot.

mean AE score of APP in a stratified validation split consisting of 40% of the training set. The model, with optimised hyperparameters, is re-fit on the whole training set before estimating the test prevalence values. Except for SVM(AE), that natively uses SVM^{perf} (Joachims, 2005), all other quantifiers rely on a Logistic

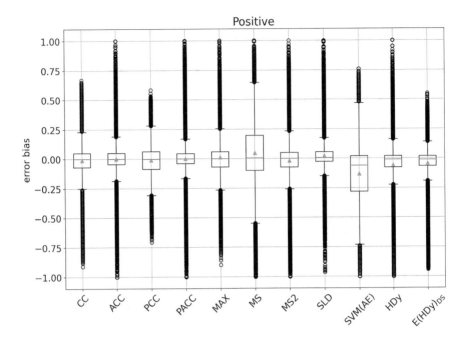

Fig. 6.5 Global Bias-Box plot.

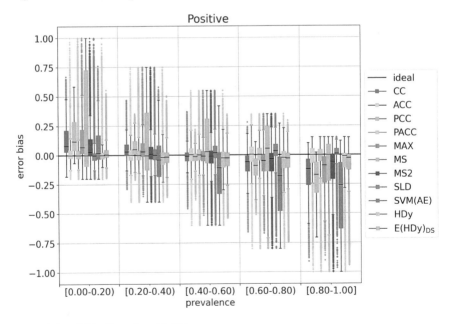

Fig. 6.6 Local-Bias-Box plot with 5 bins.

Regressor as the underlying classifier. We explore the regularisation parameter C (common to LR and SVM) in $\{10^{-3}, 10^{-2}, \ldots, 10^{2}, 10^{3}\}$, and the parameter `class_weight` (only for LR) in {"balanced", "not balanced"}.

These results are fairly consistent with other results previously reported in the literature (Moreo and Sebastiani, 2021, 2022; Pérez-Gállego et al., 2019; Schumacher et al., 2021). They clearly indicate the quantifier SLD behaves very well overall (in this case beating all other methods in 13 datasets out of 30). Methods E(HDy)$_{DS}$ (8 times best method), PACC (4 times best method), and (to a lesser extent) ACC (2 times best method), also fare very well, obtaining average ranks not statistically significantly different from the best average rank obtained by SLD. The method SVM(AE) tends to produce results that are markedly worse than the rest of competitors. In line with the observations of Schumacher et al. (2021), none of the variants MAX, MS, MS2 improve over ACC. Also in line with the findings of Pérez-Gállego et al. (2019), the ensemble E(HDy)$_{DS}$ clearly outperforms the base quantifier HDy it is built upon. A general trend that emerges in this experimentation, and that is consistent with almost any other (not to say all) reported experiments, concerns the fact that performing classification alone (as, e.g., CC, PCC, SVM(AE)) does not suffice for providing accurate estimations of class prevalence values in situations of distribution shift; in such situations one typically needs to perform some sort of adjustment to the prevalence estimation derived from the use of a (biased) classifier.

6.3.2 Visualisation Tools for the Analysis of Results

While averaged error scores do certainly speak clearly about the macro behaviour of quantification systems, they do not tell the entire story. The analysis of results can sometimes be complemented with the aid of visualisation tools that can help to unravel how a system performs in specific experimental conditions. This is specially useful in scenarios in which the practitioner wants to better understand how the system fares, say, in presence of high/low shift, or in regions of high/low prevalence. Complementing the analysis with such additional viewpoints is interesting since, for reasons discussed in Section 3.4.4, some protocols are sometimes criticised for involving testing conditions that some practitioners might deem unlikely to occur in real cases. In what follows, we discuss some useful types of plots that can be helpful in practical scenarios.

One plot which is of particular relevance for the analysis of binary quantifiers is the so-called "diagonal plot". This plot displays the predicted prevalence values along the y-axis against the true prevalence values in the x-axis; predicted values are sometimes binned according to the true prevalence values. The plot is called "diagonal" since the ideal quantifier is described by a diagonal line, from coordinates (0,0) to (1,1). An example of this plot, computed on the same batch of experiments reported in Table 6.2, is offered in Figure 6.3 (we also showed some examples in Section 1.6). This type of plot allows one to rapidly grasp intuitions about the tendency of a quantification method to systematically overestimate or underestimate the true class prevalence. In this example, the plot reveals that, for high prevalence values of the **Positive** class, SLD tends to slightly overestimate the class prevalence values, while most other methods tend instead to underestimate them. For low prevalence values of the **Positive** class, methods MAX, MS, MS2, PCC, and CC show a tendency to overestimate these prevalence values.

This plot is sometimes enriched by error bars, or colour bands around the averaged results, representing the deviations from the average. It is, however, sometimes cumbersome to plot all this information in a single plot, with many graphical elements ending up inevitably juxtaposed on top of each other. Not displaying them might however lead to misleading conclusions, since a method displaying high variance could anyway seem to perform very well, by looking at the averages of predictions, whenever the estimator is an unbiased one. (In this example, we have opted for omitting them for the sake of clarity, but some examples of diagonal plots including colour bands can be found in Esuli et al. (2018), or in the Figures 1.5 and 1.6 accompanying this book.) Yet another limitation of this kind of plot is that it is reserved for binary problems only. While it is true that one could display a dedicated diagonal plot for each of the classes in a SLQ problem, it is no less true that the intuitions one gains by inspecting diagonal plots get blurred as the number of classes increase.

Another type of plot that does not present this limitation is what we might call the "Error-by-Shift plot". This plot displays any target error metric (say, AE) along the y-axis as a function of the distribution shift between the training set and each of the test samples, on the x-axis. As for the diagonal plot, one typically displays averaged values across bins; here too, error bars or colour bands might help to reveal the system variance provided that the number of visual elements is moderate. Since this plot works with the concept of "shift" (as implemented in terms of any error or divergence metric), it can be applied to any problem characterised by any number of classes. Figure 6.5 shows an example for the experiments of Table 6.2. Note that the errors in the left-hand size of the plot correspond to situations in which the test and the training prevalence are close to each other, while errors on the right-hand size of the plot describe how the systems perform in cases of high shift. In this particular example, the plot reveals how $E(HDy)_{DS}$ excels at situations characterised by low distribution shift, while SLD seems the most robust in dealing with high-shift scenarios. This example consists of averages across 30 datasets, and so for many of them there are few, or none at all, cases of very high shift; this explains why the curves look less stable in the right-most part of the plot.

As mentioned before, both the Diagonal plot and the Error-by-Shift plot struggle to display error variances when the number of methods to compare becomes relatively high. The "Bias-Box plot" is specifically devised for studying the distribution of the error predictions in such cases. This kind of plot resorts to the well-known box plots to display the bias of the system, i.e., the signed error difference between the true class prevalence and the predicted class prevalence (see Equation 3.1 in Section 3.1.2). A box plot summarises a distribution by means of different graphical elements: the extremes of the box delimit the first and third quartiles of a distribution, a central line represents the median of the distribution while the position of a small triangle represents the average of the distribution, the maximum and the minimum are represented by the whiskers on the top and on the bottom the box, and finally the outliers appear above or below the corresponding whiskers. Figure 6.5 shows the Bias-Box plot of our experiments. This diagram reveals that PACC, SLD, and $E(HDy)_{DS}$ are the methods displaying the lowest bias overall, given that their boxes are the most squashed, and given that their whiskers are the shortest. Note how the reduction of variance with respect to the base members (HDy) that characterise the ensemble methods ($E(HDy)_{DS}$) is clearly perceivable in the last two boxes; this is in line with the observations reported by the inventors of this method (Pérez-Gállego et al., 2019). It is also interesting to note how the heuristic implemented in MS2 drastically reduces the variance displayed by MS.

As in the other cases, this plot is not exempt from limitations, though. Given that this plot uses distributions based on the bias (signed error difference), this plot gets unavoidably tied to one class (acting as the positive class), and is thus more appealing for binary problems. Yet another limitation of the Bias-Box plot has to do with the fact that the distribution of the bias is computed on the whole experiment, which might involve (as it does indeed involve when APP is adopted) cases of severe distribution shift mixed up with cases of very low shift. The "Local-Bias-Box plot" can be of help in situations in which one prefers to crumble up the distribution in different pieces each characterised by a different prevalence range, or by a different range of shift. In Figure 6.6 we show the Local-Bias-Box plot for our experiments, in which we bin the error bias in five ranges of true prevalence. This plot reveals how the "unadjusted" methods (e.g., CC, PCC) display positive bias for low prevalence values (thus showing a tendency to overestimate the true prevalence) and negative bias for high prevalence values (thus showing a tendency to underestimate the true prevalence). The "adjusted" versions (ACC and PACC), on the contrary, manage to reduce this effect, as witnessed by the fact that their box plots are centred at zero bias in those cases. This plot also reveals that MS tends to display a huge positive bias in the low-prevalence regime, while SVM(AE) displays a huge negative bias in the high-prevalence regime.

6.4   Related Tasks

6.4.1   Links to Existing Tasks

Quantification bears strong relations with *prevalence estimation from screening tests*, an important task in epidemiology (see Levy and Kass, 1970; Lew and Levy, 1989; Rahme and Joseph, 1998; Zhou et al., 2002); indeed, as already hinted in Section 6.1, the ACC quantification method discussed in Section 4.2.3 was used (in its binary form) for this task well before research in quantification was born. A screening test is a test that a patient undergoes in order to check if s/he has a given pathology. Tests are often imperfect, i.e., they may give rise to false positives (the patient is incorrectly diagnosed with the pathology) and false negatives (the test wrongly diagnoses the patient to be free from the pathology). Therefore, testing a patient is akin to classifying a data item, and using these tests for estimating the prevalence of the pathology in a given population is akin to performing quantification via classification. The main difference between this task and quantification is that a screening test typically has known and fairly constant recall (that epidemiologists call "sensitivity") and fallout (whose complement epidemiologists call "specificity"), while the same usually does not happen for a classifier.

Quantification is also closely related to the problem of *density estimation* (Silverman, 1986), defined as the estimation, based on observed data, of the unknown probability density function of a given random variable; if the random variable is discrete, this means estimating, using observed data, the unknown distribution across the discrete set of events, i.e., across the classes. A classic, textbook example of density estimation is estimating the prevalence of white balls in a large urn containing white balls and black balls. However, quantification and density estimation are different in at least two respects. First, the above "urn" example assumes that, when we pick a ball from the urn, we can deterministically assess whether the ball is black or white, by simple visual inspection; in quantification we instead assume that assessing whether a given item belongs to the class is not a deterministic operation, and depends on subjective judgment. A second key difference is that the density estimation problem arises from the fact that in many applications it is practically impossible to assess class membership for each single individual (e.g., we do *not* want to inspect every single ball in the urn); however, in the case of quantification it is feasible to analyse every single item, since this is done automatically. (This is due to the fact that the items that are the object of quantification are digital objects, and any number of them can be processed given enough computational resources.) These differences clearly indicate the existence of a task different from density estimation, and characterised (a) by the need to assess class prevalence when class membership cannot be established deterministically, and (b) by the fact that *all* individuals contained in the sample can be analysed. These facts indicate altogether that our task is closely related to *classification*, a task in which facts (a) and (b) both hold. However, the goal of classification is different from the one we have set ourselves, since in classification we are interested in correctly estimating the true class of each single item.

A research area that might seem related to quantification is *collective classification* (CoC) (Sen et al., 2008), as in statistical relational learning. Similarly to quantification, in CoC the classification of instances is not viewed in isolation. However, CoC is radically different from quantification in that its focus is on improving the accuracy of classification by exploiting relationships between the items to classify (e.g., hypertextual documents that link to each other). For instance, in certain applications characterised by "homophily" (i.e., the tendency of individuals to associate with their similar) the fact that a data item has a certain label may provide additional evidence towards the fact that a related data item (say, one that is hyperlinked to the previous one) may have that label too. Differently from quantification, CoC assumes the existence of explicit relationships between the items to classify (which quantification does not), and is evaluated at the individual level, rather than at the aggregate level as quantification is.

Another related research task is *divergence approximation* (Sugiyama et al., 2013), which consists of estimating the divergence between two distributions. This seems, on the surface, akin to evaluating the accuracy of quantification. However, the main difference is that divergence approximation is performed when one does not have access to the two distributions, but only to finite samples from them. In other words, divergence approximation is useful when one is interested in the divergence of two distributions that should be estimated via the density estimation techniques previously discussed in this section: in this case, as Sugiyama et al. (2013) put it, "directly approximating the divergence without estimating probability distributions is more sensible than a naive two-step approach of first estimating probability distributions and then approximating the divergence." Evaluating the accuracy of quantification is thus different from divergence approximation because of the very same factors that make quantification and density estimation different.

Yet another related task is *learning with label proportions* (de Freitas and Kück, 2005; Quadrianto et al., 2009), which consists of learning to estimate the class labels of individual items when training data comes in the form of samples of such items with labels at the aggregate level. In other words, we do not know the class labels of individual training items, but we only know the class prevalence of samples of such items. This is the other way around with respect to quantification, where we need to predict labels at the aggregate level by learning from training data which are labelled at the individual level.

6.4.2   A Possible Variant of the Quantification Task

Quantification, as defined in this book and in the literature that this book looks at, is (somehow similarly to learning with label proportions) an unusual supervised learning task, in that the labels that we need to predict and the labels we use in order to train our predictors are not homologous, i.e., are not of the same type. In fact, in quantification we start from a training set of labelled items, and we need to predict the prevalence of the classes in a sample of unlabelled items. In other words, in the training data the labels (i) are attached to each individual item, and (ii) are drawn from the set $\mathcal{Y}$ of classes, while in the unlabelled data for which we need to issue predictions the labels (iii) must be attached to each pair consisting of a sample (i.e., a *set* of individual items) and a class, and (iv) are drawn from the [0,1] interval. This is unlike most other tasks in supervised learning (e.g., classification, regression), where the training items and the unlabelled items that need to be labelled are homologous, and where the labels of the training items and the labels to be attached to the unlabelled items are drawn from the same set.

In the future, one might want to investigate a variant of the quantification task in which the training data and the unlabelled data are homologous, and where the training labels and the labels to be predicted are homologous too. In this variant, the training data would thus consist not of a set of items labelled at the individual level, but of a set of labelled *samples*, where the labels are from the [0,1] interval and where no labels are attached to the individual items. The advantage of this formulation would be the possibility to use more standard tools from the arsenal of supervised learning machinery, since this would squarely be a standard regression task (albeit one in which the label $p_\sigma(y_i)$ for a pair (σ, y_i) must be in [0,1] and the sum $\sum_{y_i \in \mathcal{Y}} p_\sigma(y_i)$ must be equal to 1).[4]

The disadvantage of this formulation is that it may appear unnatural, since in many applications labelled data tend to come in the form of labelled individual items, rather than labelled samples. Still, applications in which there is no access to the individual labels but a label at the collective level is available, indeed exist (as in the "learning with label proportions" task); for instance, in datasets of a medical nature the individual labels of training data might be masked off due to privacy considerations, but a label for the entire set might be available. In the future it might be interesting to investigate whether the advantages brought about by this formulation offset its disadvantages or not.

---

[4] Indeed, this formulation is clearly reminiscent of *multiple-instance regression* (Dooly et al., 2002; Ray and Page, 2001), a class of supervised learning techniques in which individual items (called *instances*) each have a vectorial representation and are grouped into sets (called *bags*). In multiple-instance regression, only the bags, and not the individual instances, have (real-valued) labels.

Chapter 7
The Road Ahead

Quantification has seen a growing amount of work in the last 15 years, spawned by the realisation that there are a lot of application settings in which the class labels to be attributed to individual items are not interesting *per se*, but are only the stepping stones towards estimating prevalence values for the classes of interest. While research on learning to quantify has grown steadily since 2005 onwards, much more is still needed in order to stably deliver accurate results across the entire range of applicative settings on which quantification can be employed.

What is the road ahead, then, for learning to quantify? While there are margins of improvements on all the areas that this book has touched upon, from plain single-label quantification to the more complex ordinal quantification, from standard application contexts to more peculiar ones involving, say, streaming data or multilingual text, we think there are a few "burning topics" which are sorely in need of (and that are likely to see) further work:

- *Quantification and deep learning.* While deep learning has had an enormous impact on AI and machine learning in general, and on classification in particular, there has not been much work on applying deep learning to quantification; so far, the only works in this department are Esuli et al. (2018), Sanya et al. (2018), and Qi et al. (2020), discussed in Sections 4.2.12 and 4.3.1, respectively. While nowadays neural architectures naturally cater for variable-length sequential data, how to properly represent (or embed) unordered sets of elements is less clear, and it has been shown that simply arranging the elements in the set in an arbitrary order is problematic (Vinyals et al., 2016). Since unordered sets represent the primary form of interest in learning to quantify, it is likely that the study of permutation-invariant functions will become a central subject in future research on deep learning and quantification. Although some attempts have been made in trying to represent unordered sets of inputs with deep learning architectures (Vinyals et al., 2016; Zaheer et al., 2017), more recent work suggests that this field is yet to be well understood (Wagstaff et al., 2019).

© The Author(s) 2023
A. Esuli et al., *Learning to Quantify*, The Information Retrieval Series 47,
https://doi.org/10.1007/978-3-031-20467-8_7

- *Non-aggregative methods for learning to quantify.* Most approaches to learning to quantify that have been proposed up to now consist of aggregative methods, while research on non-aggregative methods has been somehow lagging behind. It is our impression that the future of learning to quantify will be in the non-aggregative camp, though, since these methods are, as observed at the beginning of Section 4.4, the true realisation of Vapnik's principle as applied to quantification; it is by fully taking advantage of the fact that quantification is an "easier" problem than classification, that quantification accuracy will witness a substantial improvement.

- *Explainable quantification.* Machine learning algorithms have traditionally been opaque. In recent years, though, driven by the need to ensure fairness and transparency in the decision-making processes that ML-powered algorithms allow, we have seen a surge of interest in making the inferences carried out by these algorithms (e.g., the classification decisions) explainable. While explainability has been a hot issue for classification, we are not aware of any work yet on explainable quantification. One problem that researchers who will address this area should tackle, is the fact that, if we train our best classifier and our best quantifier, their decisions will in general not coincide (unless we use Classify and Count); in other words, if we classify all the unlabelled items via our classifier and count the items that have been assigned a certain class, the resulting prevalence value will not coincide with the one estimated by our quantifier. Providing these results, say, to a customer that has commissioned us with this work, may prove embarrassing, and hard to convincingly explain in layman's terms.

  One possible solution might consist of ranking the unlabelled items in decreasing order of the posterior probabilities generated by our classifier, and setting the classification threshold exactly at the value that justifies the class prevalence estimated by our quantifier; the items ranked above the threshold would thus constitute the "explanation" of the class prevalence returned by our quantifier. (The classifier, if generated with "explainable machine learning" technology, would in turn provide explanations for its individual classification decisions.) Still, this threshold would be different from the one that "our best possible classifier" would use, which makes this solution suboptimal. Research on quantification and explainability is thus sorely needed.

- *Transductive quantification.* A number of applications of quantification are transductive in nature, i.e., there is a single, finite set of unlabelled items for which we are interested in estimating class prevalence values, and this set is available at training time. For instance, in the "What do you think of onions in cheeseburgers?" scenario mentioned at the very beginning of Chapter 1, the market research expert may be interested in running this survey monthly, in order to track the evolution of customers' preferences (such a survey would be called a "tracker", in market research jargon). Alternatively, she might be interested in running the survey only once, in a one-off manner; in this case, the quantifier can be trained "on purpose" once the survey data are in, and the training process can take advantage from the fact that the data to quantify on are already available.

Transductive quantification is yet another context in which Vapnik's principle applies: estimating class prevalence values for a finite set of data is a less general (hence simpler) problem than generating a quantifier that generalises to the entire domain. So far, this aspect has been exploited by a few methods, e.g., in Saerens et al. (2002)'s SLD method (Section 4.2.9) and Xue and Weiss (2009)'s CDE-Iterate method (Section 4.2.10); the fact that, for tasks other than quantification, transductive inference has been investigated quite frequently in recent years, and the abundance of contexts to which it can be applied, should incentivise researchers in devoting more effort to this area.

However, if there is one aspect of the quantification task that is even more sorely in need of advancement than the ones mentioned above, this is the awareness of its very existence on the part of its potential users. The large majority of application papers in which class prevalence values need to be estimated on sets of unlabelled data, still use Classify and Count, essentially because the authors ignore that there is a better alternative out there. Raising the awareness that class prevalence estimation is a problem that should be solved by its own specific techniques is a necessary step. This awareness is important especially since, with the advent of big data, more and more application contexts spring up in which we cannot afford analysing the data at the individual level, and the aggregate level is what we have to be happy with.

Bibliography

Alaíz-Rodríguez, R., Guerrero-Curieses, A., and Cid-Sueiro, J. (2011). Class and subclass probability re-estimation to adapt a classifier in the presence of concept drift. *Neurocomputing*, 74(16):2614–2623.

Alexandari, A., Kundaje, A., and Shrikumar, A. (2020). Maximum likelihood with bias-corrected calibration is hard-to-beat at label shift adaptation. In *Proceedings of the 37th International Conference on Machine Learning (ICML 2020)*, pages 222–232, Vienna, AT.

Anderson, T. W. (1962). On the distribution of the two-sample Cramer-von Mises criterion. *The Annals of Mathematical Statistics*, 33(3):1148–1159.

Andrus, M., Spitzer, E., Brown, J., and Xiang, A. (2021). What we can't measure, we can't understand: Challenges to demographic data procurement in the pursuit of fairness. In *Proceedings of the 4th ACM Conference on Fairness, Accountability, and Transparency (FAccT 2021)*, pages 249–260, Toronto, CA.

Arribas, J. I. and Cid-Sueiro, J. (2005). A model selection algorithm for a posteriori probability estimation with neural networks. *IEEE Transactions on Neural Networks*, 16(4):799–809.

Baccianella, S., Esuli, A., and Sebastiani, F. (2013). Variable-constraint classification and quantification of radiology reports under the ACR Index. *Expert Systems and Applications*, 40(9):3441–3449.

Balikas, G., Partalas, I., Gaussier, E., Babbar, R., and Amini, M.-R. (2015). Efficient model selection for regularized classification by exploiting unlabeled data. In *Proceedings of the 14th International Symposium on Intelligent Data Analysis (IDA 2015)*, pages 25–36, Saint Etienne, FR.

Barocas, S., Hardt, M., and Narayanan, A. (2019). *Fairness and machine learning*. fairmlbook.org.

Barranquero, J., Díez, J., and del Coz, J. J. (2015). Quantification-oriented learning based on reliable classifiers. *Pattern Recognition*, 48(2):591–604.

Barranquero, J., González, P., Díez, J., and del Coz, J. J. (2013). On the study of nearest neighbor algorithms for prevalence estimation in binary problems. *Pattern Recognition*, 46(2):472–482.

Beijbom, O., Hoffman, J., Yao, E., Darrell, T., Rodriguez-Ramirez, A., Gonzalez-Rivero, M., and Hoegh-Guldberg, O. (2015). Quantification in-the-wild: Data-sets and baselines. CoRR abs/1510.04811 (2015). Presented at the NIPS 2015 Workshop on Transfer and Multi-Task Learning, Montreal, CA.

Bella, A., Ferri, C., Hernández-Orallo, J., and Ramírez-Quintana, M. J. (2010). Quantification via probability estimators. In *Proceedings of the 11th IEEE International Conference on Data Mining (ICDM 2010)*, pages 737–742, Sydney, AU.

Bella, A., Ferri, C., Hernández-Orallo, J., and Ramírez-Quintana, M. J. (2014). Aggregative quantification for regression. *Data Mining and Knowledge Discovery*, 28(2):475–518.

© The Author(s) 2023
A. Esuli et al., *Learning to Quantify*, The Information Retrieval Series 47,
https://doi.org/10.1007/978-3-031-20467-8

Biswas, A. and Mukherjee, S. (2021). Ensuring fairness under prior probability shifts. In *Proceedings of the 2021 AAAI/ACM Conference on AI, Ethics, and Society (AIES 2021)*, pages 414–424, [Virtual Event]].

Bogen, M., Rieke, A., and Ahmed, S. (2020). Awareness in practice: Tensions in access to sensitive attribute data for antidiscrimination. In *Proceedings of the 3rd ACM Conference on Fairness, Accountability, and Transparency (FAT* 2020)*, pages 492–500, Barcelona, ES.

Borge-Holthoefer, J., Magdy, W., Darwish, K., and Weber, I. (2015). Content and network dynamics behind Egyptian political polarization on Twitter. In *Proceedings of the 18th ACM Conference on Computer Supported Cooperative Work and Social Computing (CSCW 2015)*, pages 700–711, Vancouver, CA.

Busin, L. and Mizzaro, S. (2013). Axiometrics: An axiomatic approach to information retrieval effectiveness metrics. In *Proceedings of the 4th International Conference on the Theory of Information Retrieval (ICTIR 2013)*, page 8, Copenhagen, DK.

Calders, T. and Verwer, S. (2010). Three naive Bayes approaches for discrimination-free classification. *Data Mining and Knowledge Discovery*, 21(2):277–292.

Card, D. and Smith, N. A. (2018). The importance of calibration for estimating proportions from annotations. In *Proceedings of the 2018 Conference of the North American Chapter of the Association for Computational Linguistics (HLT-NAACL 2018)*, pages 1636–1646, New Orleans, US.

Ceron, A., Curini, L., and Iacus, S. M. (2016). iSA: A fast, scalable and accurate algorithm for sentiment analysis of social media content. *Information Sciences*, 367/368:105—124.

Ceron, A., Curini, L., Iacus, S. M., and Porro, G. (2014). Every tweet counts? How sentiment analysis of social media can improve our knowledge of citizens' political preferences with an application to Italy and France. *New Media & Society*, 16(2):340–358.

Chakrabarti, S., Dom, B. E., and Indyk, P. (1998). Enhanced hypertext categorization using hyperlinks. In *Proceedings of the 24th ACM International Conference on Management of Data (SIGMOD 1998)*, pages 307–318, Seattle, US.

Chan, Y. S. and Ng, H. T. (2005). Word sense disambiguation with distribution estimation. In *Proceedings of the 19th International Joint Conference on Artificial Intelligence (IJCAI 2005)*, pages 1010–1015, Edinburgh, UK.

Chan, Y. S. and Ng, H. T. (2006). Estimating class priors in domain adaptation for word sense disambiguation. In *Proceedings of the 44th Annual Meeting of the Association for Computational Linguistics (ACL 2006)*, pages 89–96, Sydney, AU.

Criminisi, A., Shotton, J., and Konukoglu, E. (2011). Decision forests: A unified framework for classification, regression, density estimation, manifold learning and semi-supervised learning. *Foundations and Trends in Computer Graphics and Vision*, 7(2/3):81–227.

Da San Martino, G., Gao, W., and Sebastiani, F. (2016a). Ordinal text quantification. In *Proceedings of the 39th ACM Conference on Research and Development in Information Retrieval (SIGIR 2016)*, pages 937–940, Pisa, IT.

Da San Martino, G., Gao, W., and Sebastiani, F. (2016b). QCRI at SemEval-2016 Task 4: Probabilistic methods for binary and ordinal quantification. In *Proceedings of the 10th International Workshop on Semantic Evaluation (SemEval 2016)*, pages 58–63, San Diego, US.

Daughton, A. R. and Paul, M. J. (2019). Constructing accurate confidence intervals when aggregating social media data for public health monitoring. In *Proceedings of the 3rd AAAI International Workshop on Health Intelligence (W3PHIAI 2019)*, pages 9–17, Phoenix, US.

de Freitas, N. and Kück, H. (2005). Learning about individuals from group statistics. In *Proceedings of the 21st Conference in Uncertainty in Artificial Intelligence (UAI 2005)*, pages 332–339, Edimburgh, UK.

Dempster, A. P., Laird, N. M., and Rubin, D. B. (1977). Maximum likelihood from incomplete data via the EM algorithm. *Journal of the Royal Statistical Society, B*, 39(1):1–38.

Denham, B., Lai, E. M., Sinha, R., and Naeem, M. A. (2021). Gain-Some-Lose-Some: Reliable quantification under general dataset shift. In *Proceedings of the 2021 IEEE International Conference on Data Mining (ICDM 2021)*, pages 1048–1053.

Dodds, P. S., Harris, K. D., Kloumann, I. M., Bliss, C. A., and Danforth, C. M. (2011). Temporal patterns of happiness and information in a global social network: Hedonometrics and Twitter. *PLoS ONE*, 6(12):1–26.

Domingos, P. M. and Pazzani, M. J. (1997). On the optimality of the simple Bayesian classifier under zero-one loss. *Machine Learning*, 29(2-3):103–130.

Dooly, D. R., Zhang, Q., Goldman, S. A., and Amar, R. A. (2002). Multiple-instance learning of real-valued data. *Journal of Machine Learning Research*, 3:651–678.

dos Reis, D. M., Flach, P., Matwin, S., and Batista, G. (2016). Fast unsupervised online drift detection using incremental Kolmogorov-Smirnov test. In *Proceedings of the 22nd ACM SIGKDD International Conference on Knowledge Discovery and Data Mining (KDD 2016)*, pages 1545–1554, San Francisco, US.

du Plessis, M. C., Niu, G., and Sugiyama, M. (2017). Class-prior estimation for learning from positive and unlabeled data. *Machine Learning*, 106(4):463–492.

du Plessis, M. C. and Sugiyama, M. (2012). Semi-supervised learning of class balance under class-prior change by distribution matching. In *Proceedings of the 29th International Conference on Machine Learning (ICML 2012)*, Edinburgh, UK.

du Plessis, M. C. and Sugiyama, M. (2014). Class prior estimation from positive and unlabeled data. *IEICE Transactions*, 97-D(5):1358–1362.

Duda, R. O., Hart, P. E., and Stork, D. G. (2001). *Pattern classification*. John Wiley & Sons, New York, US, 2nd edition.

Elkan, C. (2001). The foundations of cost-sensitive learning. In *Proceedings of the 17th International Joint Conference on Artificial Intelligence (IJCAI 2001)*, pages 973–978, Seattle, US.

Elliott, M. N., Morrison, P. A., Fremont, A., McCaffrey, D. F., Pantoja, P., and Lurie, N. (2009). Using the Census Bureau's surname list to improve estimates of race/ethnicity and associated disparities. *Health Services and Outcomes Research Methodology*, 9(2):69–83.

Esuli, A. (2016). ISTI-CNR at SemEval-2016 Task 4: Quantification on an ordinal scale. In *Proceedings of the 10th International Workshop on Semantic Evaluation (SemEval 2016)*, San Diego, US.

Esuli, A., Molinari, A., and Sebastiani, F. (2021). A critical reassessment of the Saerens-Latinne-Decaestecker algorithm for posterior probability adjustment. *ACM Transactions on Information Systems*, 39(2):Article 19.

Esuli, A., Moreo, A., and Sebastiani, F. (2018). A recurrent neural network for sentiment quantification. In *Proceedings of the 27th ACM International Conference on Information and Knowledge Management (CIKM 2018)*, pages 1775–1778, Torino, IT.

Esuli, A., Moreo, A., and Sebastiani, F. (2020). Cross-lingual sentiment quantification. *IEEE Intelligent Systems*, 35(3):106–114.

Esuli, A., Moreo, A., and Sebastiani, F. (2022). LeQua@CLEF2022: Learning to Quantify. In *Proceedings of the 44th European Conference on Information Retrieval (ECIR 2022)*, pages 374–381, Stavanger, NO.

Esuli, A. and Sebastiani, F. (2010a). Machines that learn how to code open-ended survey data. *International Journal of Market Research*, 52(6):775–800.

Esuli, A. and Sebastiani, F. (2010b). Sentiment quantification. *IEEE Intelligent Systems*, 25(4):72–75.

Esuli, A. and Sebastiani, F. (2013). Improving text classification accuracy by training label cleaning. *ACM Transactions on Information Systems*, 31(4):Article 19.

Esuli, A. and Sebastiani, F. (2014). Explicit loss minimization in quantification applications (preliminary draft). In *Proceedings of the 8th International Workshop on Information Filtering and Retrieval (DART 2014)*, pages 1–11, Pisa, IT.

Esuli, A. and Sebastiani, F. (2015). Optimizing text quantifiers for multivariate loss functions. *ACM Transactions on Knowledge Discovery and Data*, 9(4):Article 27.

Fabris, A., Esuli, A., Moreo, A., and Sebastiani, F. (2021). Measuring fairness under unawareness via quantification. *arXiv preprint arXiv:2109.08549*.

Fawcett, T. and Flach, P. (2005). A response to Webb and Ting's 'On the application of ROC analysis to predict classification performance under varying class distributions'. *Machine Learning*, 58(1):33–38.

Fernandes Vaz, A., Izbicki, R., and Bassi Stern, R. (2017). Prior shift using the ratio estimator. In *Proceedings of the International Workshop on Bayesian Inference and Maximum Entropy Methods in Science and Engineering*, pages 25–35, Jarinu, BR.

Fernandes Vaz, A., Izbicki, R., and Bassi Stern, R. (2019). Quantification under prior probability shift: The ratio estimator and its extensions. *Journal of Machine Learning Research*, 20:79:1–79:33.

Flach, P. A. (2017). Classifier calibration. In Sammut, C. and Webb, G. I., editors, *Encyclopedia of Machine Learning*, pages 212–219. Springer, Heidelberg, DE, 2nd edition.

Forman, G. (2005). Counting positives accurately despite inaccurate classification. In *Proceedings of the 16th European Conference on Machine Learning (ECML 2005)*, pages 564–575, Porto, PT.

Forman, G. (2006). Quantifying trends accurately despite classifier error and class imbalance. In *Proceedings of the 12th ACM SIGKDD International Conference on Knowledge Discovery and Data Mining (KDD 2006)*, pages 157–166, Philadelphia, US.

Forman, G. (2008). Quantifying counts and costs via classification. *Data Mining and Knowledge Discovery*, 17(2):164–206.

Forman, G., Kirshenbaum, E., and Suermondt, J. (2006). Pragmatic text mining: Minimizing human effort to quantify many issues in call logs. In *Proceedings of the 12th ACM International Conference on Knowledge Discovery and Data Mining (KDD 2006)*, pages 852–861, Philadelphia, US.

Gao, W. and Sebastiani, F. (2015). Tweet sentiment: From classification to quantification. In *Proceedings of the 7th International Conference on Advances in Social Network Analysis and Mining (ASONAM 2015)*, pages 97–104, Paris, FR.

Gao, W. and Sebastiani, F. (2016). From classification to quantification in tweet sentiment analysis. *Social Network Analysis and Mining*, 6(19):1–22.

Gart, J. J. and Buck, A. A. (1966). Comparison of a screening test and a reference test in epidemiologic studies: II. A probabilistic model for the comparison of diagnostic tests. *American Journal of Epidemiology*, 83(3):593–602.

González, P., Álvarez, E., Díez, J., López-Urrutia, A., and del Coz, J. J. (2017). Validation methods for plankton image classification systems. *Limnology and Oceanography: Methods*, 15:221–237.

González, P., Díez, J., Chawla, N., and del Coz, J. J. (2017). Why is quantification an interesting learning problem? *Progress in Artificial Intelligence*, 6(1):53–58.

González-Castro, V., Alaiz-Rodríguez, R., and Alegre, E. (2013). Class distribution estimation based on the Hellinger distance. *Information Sciences*, 218:146–164.

González-Castro, V., Alaiz-Rodríguez, R., Fernández-Robles, L., Guzmán-Martínez, R., and Alegre, E. (2010). Estimating class proportions in boar semen analysis using the Hellinger distance. In *Proceedings of the 23rd International Conference on Industrial Engineering and other Applications of Applied Intelligent Systems (IEA/AIE 2010)*, pages 284–293, Cordoba, ES.

Grimmer, J., Messing, S., and Westwood, S. J. (2012). How words and money cultivate a personal vote: The effect of legislator credit claiming on constituent credit allocation. *American Political Science Review*, 106(4):703–719.

Guerrero-Curieses, A., Alaiz-Rodríguez, R., and Cid-Sueiro, J. (2005). Loss functions to combine learning and decision in multiclass problems. *Neurocomputing*, 69(1-3):3–17.

Hand, D. J. and Henley, W. E. (1997). Statistical classification methods in consumer credit scoring: A review. *Journal of the Royal Statistical Society: Series A (Statistics in Society)*, 160(3):523–541.

Hassan, W., Maletzke, A., and Batista, G. (2020). Accurately quantifying a billion instances per second. In *Proceedings of the 7th IEEE International Conference on Data Science and Advanced Analytics (DSAA 2020)*, pages 1–10, Sydney, AU.

Hassan, W., Maletzke, A. G., and Batista, G. (2021). Pitfalls in quantification assessment. In Cong, G. and Ramanath, M., editors, *Proceedings of the CIKM 2021 Workshops co-located with 30th ACM International Conference on Information and Knowledge Management (CIKM 2021), Gold Coast, Queensland, Australia, November 1-5, 2021*, volume 3052 of *CEUR Workshop Proceedings*. CEUR-WS.org.

Holstein, K., Wortman Vaughan, J., Daumé III, H., Dudik, M., and Wallach, H. (2019). Improving fairness in machine learning systems: What do industry practitioners need? In *Proceedings of the ACM Conference on Human Factors in Computing Systems (CHI 2019)*, pages 1–16, Glasgow, UK.

Hopkins, D. J. and King, G. (2010). A method of automated nonparametric content analysis for social science. *American Journal of Political Science*, 54(1):229–247.

Iyer, A., Nath, S., and Sarawagi, S. (2014). Maximum mean discrepancy for class ratio estimation: Convergence bounds and kernel selection. In *Proceedings of the 31st International Conference on Machine Learning (ICML 2014)*, pages 530–538, Beijing, CN.

Jerzak, C. T., King, G., and Strezhnev, A. (2022). An improved method of automated nonparametric content analysis for social science. *Political Analysis*. Forthcoming.

Joachims, T. (1999). Transductive inference for text classification using support vector machines. In *Proceedings of the 16th International Conference on Machine Learning (ICML 1999)*, pages 200–209, Bled, SL.

Joachims, T. (2005). A support vector method for multivariate performance measures. In *Proceedings of the 22nd International Conference on Machine Learning (ICML 2005)*, pages 377–384, Bonn, DE.

Johnson, D. and Sinanovic, S. (2001). Symmetrizing the Kullback-Leibler distance. *IEEE Transactions on Information Theory*, 1(1):1–10.

Kar, P., Li, S., Narasimhan, H., Chawla, S., and Sebastiani, F. (2016). Online optimization methods for the quantification problem. In *Proceedings of the 22nd ACM SIGKDD International Conference on Knowledge Discovery and Data Mining (KDD 2016)*, pages 1625–1634, San Francisco, US.

Keith, K. A. and O'Connor, B. (2018). Uncertainty-aware generative models for inferring document class prevalence. In *Proceedings of the 2018 Conference on Empirical Methods in Natural Language Processing (EMNLP 2018)*, Brussels, BE.

King, G. and Lu, Y. (2008). Verbal autopsy methods with multiple causes of death. *Statistical Science*, 23(1):78–91.

King, G., Lu, Y., and Shibuya, K. (2010). Designing verbal autopsy studies. *Population Health Metrics*, 19(8).

King, G., Pan, J., and Roberts, M. E. (2013). How censorship in China allows government criticism but silences collective expression. *American Political Science Review*, 107(2):326–343.

Koppel, M., Schler, J., and Argamon, S. (2009). Computational methods in authorship attribution. *Journal of the American Society for Information Science and Technology*, 60(1):9–26.

Lang, K. (1995). Newsweeder: Learning to filter netnews. In *Proceedings of the 12th International Conference on Machine Learning (ICML 1995)*, pages 331–339, Tahoe City, US.

Latinne, P., Saerens, M., and Decaestecker, C. (2001). Adjusting the outputs of a classifier to new a priori probabilities may significantly improve classification accuracy: Evidence from a multi-class problem in remote sensing. In *Proceedings of the 18th International Conference on Machine Learning (ICML 2001)*, pages 298–305, Williamstown, US.

Levin, R. and Roitman, H. (2017). Enhanced probabilistic classify and count methods for multi-label text quantification. In *Proceedings of the 7th ACM International Conference on the Theory of Information Retrieval (ICTIR 2017)*, pages 229–232, Amsterdam, NL.

Levina, E. and Bickel, P. (2001). The Earth Mover's Distance is the Mallows distance: Some insights from statistics. In *Proceedings of the 8th International Conference on Computer Vision (ICCV 2001)*, pages 251–256, Vancouver, CA.

Levy, P. S. and Kass, E. H. (1970). A three-population model for sequential screening for bacteriuria. *American Journal of Epidemiology*, 91(2):148–154.

Lew, R. A. and Levy, P. S. (1989). Estimation of prevalence on the basis of screening tests. *Statistics in Medicine*, 8(10):1225–1230.

Lewis, D. D. (1995). Evaluating and optimizing autonomous text classification systems. In *Proceedings of the 18th ACM International Conference on Research and Development in Information Retrieval (SIGIR 1995)*, pages 246–254, Seattle, US.

Lewis, D. D. and Gale, W. A. (1994). A sequential algorithm for training text classifiers. In *Proceedings of the 17th ACM International Conference on Research and Development in Information Retrieval (SIGIR 1994)*, pages 3–12, Dublin, IE.

Lewis, D. D., Yang, Y., Rose, T. G., and Li, F. (2004). RCV1: A new benchmark collection for text categorization research. *Journal of Machine Learning Research*, 5:361–397.

Limsetto, N. and Waiyamai, K. (2011). Handling concept drift via ensemble and class distribution estimation technique. In *Proceedings of the 7th International Conference on Advanced Data Mining (ADMA 2011)*, pages 13–26, Bejing, CN.

Macskassy, S. A. and Provost, F. (2003). A simple relational classifier. In *Proceedings of the SIGKDD MultiRelational Data Mining Workshop (MRDM 2003)*, Washington, US.

Macskassy, S. A. and Provost, F. J. (2007). Classification in networked data: A toolkit and a univariate case study. *Journal of Machine Learning Research*, 8:935–983.

Makris, C., Panagis, Y., Sakkopoulos, E., and Tsakalidis, A. (2007). Category ranking for personalized search. *Data & Knowledge Engineering*, 60(1):109–125.

Maletzke, A., Moreira dos Reis, D., Cherman, E., and Batista, G. (2019). DyS: A framework for mixture models in quantification. In *Proceedings of the 33rd AAAI Conference on Artificial Intelligence (AAAI 2019)*, pages 4552–4560, Honolulu, US.

Maletzke, A. G., Moreira dos Reis, D., and Batista, G. E. (2017). Quantification in data streams: Initial results. In *Proceedings of the 2017 Brazilian Conference on Intelligent Systems (BRACIS 2017)*, pages 43–48, Uberlândia, BZ.

Maletzke, A. G., Moreira dos Reis, D., and Batista, G. E. (2018). Combining instance selection and self-training to improve data stream quantification. *Journal of the Brazilian Computer Society*, 24(12):43–48.

Mandel, B., Culotta, A., Boulahanis, J., Stark, D., Lewis, B., and Rodrigue, J. (2012). A demographic analysis of online sentiment during hurricane Irene. In *Proceedings of the NAACL/HLT Workshop on Language in Social Media*, pages 27–36, Montreal, CA.

Mehrabi, N., Morstatter, F., Saxena, N., Lerman, K., and Galstyan, A. (2019). A survey on bias and fairness in machine learning. arXiv 1908.09635.

Milli, L., Monreale, A., Rossetti, G., Giannotti, F., Pedreschi, D., and Sebastiani, F. (2013). Quantification trees. In *Proceedings of the 13th IEEE International Conference on Data Mining (ICDM 2013)*, pages 528–536, Dallas, US.

Milli, L., Monreale, A., Rossetti, G., Pedreschi, D., Giannotti, F., and Sebastiani, F. (2015). Quantification in social networks. In *Proceedings of the 2nd IEEE International Conference on Data Science and Advanced Analytics (DSAA 2015)*, Paris, FR.

Moreira dos Reis, D., Maletzke, A., Cherman, E., and Batista, G. E. (2018a). One-class quantification. In *Proceedings of the 29th European Conference on Machine Learning and Principles and Practice of Knowledge Discovery in Databases (ECML-PKDD 2018)*, pages 273–289, Dublin, IE.

Moreira dos Reis, D., Maletzke, A. G., Silva, D. F., and Batista, G. E. (2018b). Classifying and counting with recurrent contexts. In *Proceedings of the 24th ACM International Conference on Knowledge Discovery and Data Mining (KDD 2018)*, pages 1983–1992, London, UK.

Moreno-Torres, J. G., Raeder, T., Alaíz-Rodríguez, R., Chawla, N. V., and Herrera, F. (2012). A unifying view on dataset shift in classification. *Pattern Recognition*, 45(1):521–530.

Moreo, A., Esuli, A., and Sebastiani, F. (2016). Distributional correspondence indexing for cross-lingual and cross-domain sentiment classification. *Journal of Artificial Intelligence Research*, 55:131–163.

Moreo, A., Esuli, A., and Sebastiani, F. (2021a). QuaPy: A Python-based framework for quantification. In *Proceedings of the 30th ACM International Conference on Knowledge Management (CIKM 2021)*, pages 4534–4543, Gold Coast, AU.

Moreo, A., Esuli, A., and Sebastiani, F. (2021b). Word-class embeddings for multiclass text classification. *Data Mining and Knowledge Discovery*, 353(3):911–963.

Moreo, A. and Sebastiani, F. (2021). Re-assessing the "classify and count" quantification method. In *Proceedings of the 43rd European Conference on Information Retrieval (ECIR 2021)*, volume II, pages 75–91, Lucca, IT.

Moreo, A. and Sebastiani, F. (2022). Tweet sentiment quantification: An experimental re-evaluation. *PLoS ONE*, 17(9):1–23.

Morvan, J., Coste, J., Roux, C. H., Euller-Ziegler, L., Saraux, A., and Guillemin, F. (2008). Prevalence in two-phase surveys: Accuracy of screening procedure and corrected estimates. *Annals of Epidemiology*, 18(4):261–269.

Nakov, P., Farra, N., and Rosenthal, S. (2017). SemEval-2017 Task 4: Sentiment analysis in Twitter. In *Proceedings of the 11th International Workshop on Semantic Evaluation (SemEval 2017)*, Vancouver, CA.

Nakov, P., Ritter, A., Rosenthal, S., Sebastiani, F., and Stoyanov, V. (2016). SemEval-2016 Task 4: Sentiment analysis in Twitter. In *Proceedings of the 10th International Workshop on Semantic Evaluation (SemEval 2016)*, pages 1–18, San Diego, US.

Oard, D. W., Sebastiani, F., and Vinjumur, J. K. (2018). Jointly minimizing the expected costs of review for responsiveness and privilege in e-discovery. *ACM Transactions on Information Systems*, 37(1):11:1–11:35.

Pan, S. J. and Yang, Q. (2010). A survey on transfer learning. *IEEE Transactions on Knowledge and Data Engineering*, 22(10):1345–1359.

Platt, J. C. (2000). Probabilistic outputs for support vector machines and comparison to regularized likelihood methods. In Smola, A., Bartlett, P., Schölkopf, B., and Schuurmans, D., editors, *Advances in Large Margin Classifiers*, pages 61–74. The MIT Press, Cambridge, MA.

Prettenhofer, P. and Stein, B. (2011). Cross-lingual adaptation using structural correspondence learning. *ACM Transactions on Intelligent Systems and Technology*, 3(1):Article 13.

Pérez-Gállego, P., Castaño, A., Quevedo, J. R., and del Coz, J. J. (2019). Dynamic ensemble selection for quantification tasks. *Information Fusion*, 45:1–15.

Pérez-Gállego, P., Quevedo, J. R., and del Coz, J. J. (2017). Using ensembles for problems with characterizable changes in data distribution: A case study on quantification. *Information Fusion*, 34:87–100.

Qi, L., Khaleel, M., Tavanapong, W., Sukul, A., and Peterson, D. (2020). A framework for deep quantification learning. In *Proceedings of the European Conference on Machine Learning and Principles of Knowledge Discovery in Databases (ECML/PKDD 2020)*, pages 232–248, Ghent, BE.

Quadrianto, N., Smola, A. J., Caetano, T. S., and Le, Q. V. (2009). Estimating labels from label proportions. *Journal of Machine Learning Research*, 10:2349–2374.

Quiñonero-Candela, J., Sugiyama, M., Schwaighofer, A., and Lawrence, N. D., editors (2009). *Dataset shift in machine learning*. The MIT Press, Cambridge, US.

Rahme, E. and Joseph, L. (1998). Estimating the prevalence of a rare disease: Adjusted maximum likelihood. *The Statistician*, 47:149–158.

Ray, S. and Page, D. (2001). Multiple instance regression. In *Proceedings of the 18th International Conference on Machine Learning*, ICML 2001, pages 425–432, Williams College, US.

Rüschendorf, L. (2001). Wasserstein metric. In Hazewinkel, M., editor, *Encyclopaedia of Mathematics*. Kluwer Academic Publishers, Dordrecht, NL.

Rubner, Y., Tomasi, C., and Guibas, L. J. (1998). A metric for distributions with applications to image databases. In *Proceedings of the 6th International Conference on Computer Vision (ICCV 1998)*, pages 59–66, Mumbai, IN.

Saerens, M., Latinne, P., and Decaestecker, C. (2002). Adjusting the outputs of a classifier to new a priori probabilities: A simple procedure. *Neural Computation*, 14(1):21–41.

Sakai, T. (2018). Comparing two binned probability distributions for information access evaluation. In *Proceedings of the 41st International ACM Conference on Research and Development in Information Retrieval (SIGIR 2018)*, pages 1073–1076, Ann Arbor, US.

Sakai, T. (2021). A closer look at evaluation measures for ordinal quantification. In *Proceedings of the CIKM 2021 Workshop on Learning to Quantify*, Virtual Event.

Sanya, A., Kumar, P., Kar, P., Chawla, S., and Sebastiani, F. (2018). Optimizing non-decomposable measures with deep networks. *Machine Learning*, 107(8-10):1597–1620.

Schumacher, T., Strohmaier, M., and Lemmerich, F. (2021). A comparative evaluation of quantification methods. arXiv:2103.03223.

Sebastiani, F. (2018). Market research, deep learning, and quantification. Presented at the ASC Conference on the Application of Artificial Intelligence and Machine Learning to Surveys, London, UK. http://goo.gl/JvWU7A.

Sebastiani, F. (2020). Evaluation measures for quantification: An axiomatic approach. *Information Retrieval Journal*, 23(3):255–288.

Sen, P., Namata, G., Bilgic, M., Getoor, L., Gallagher, B., and Eliassi-Rad, T. (2008). Collective classification in network data. *AI Magazine*, 29(3):93–106.

Silverman, B. W. (1986). *Density estimation for statistics and data analysis*. Chapman and Hall, London, UK.

Smith, N. A. and Tromble, R. W. (2004). Sampling uniformly from the unit simplex. Technical report, Johns Hopkins University. https://www.cs.cmu.edu/~nasmith/papers/smith+tromble.tr04.pdf.

Spence, D., Inskip, C., Quadrianto, N., and Weir, D. (2019). Quantification under class-conditional dataset shift. In *Proceedings of the 11th International Conference on Advances in Social Networks Analysis and Mining (ASONAM 2019)*, pages 528–529, Vancouver, CA.

Storkey, A. (2009). When training and test sets are different: Characterizing learning transfer. In Quiñonero-Candela, J., Sugiyama, M., Schwaighofer, A., and Lawrence, N. D., editors, *Dataset shift in machine learning*, pages 3–28. The MIT Press, Cambridge, US.

Sugiyama, M., Liu, S., du Plessis, M. C., Yamanaka, M., Yamada, M., Suzuki, T., and Kanamori, T. (2013). Direct divergence approximation between probability distributions and its applications in machine learning. *Journal of Computing Science and Engineering*, 7(2):99–111.

Tang, L., Gao, H., and Liu, H. (2010). Network quantification despite biased labels. In *Proceedings of the 8th Workshop on Mining and Learning with Graphs (MLG 2010)*, pages 147–154, Washington, US.

Tasche, D. (2016). Does quantification without adjustments work? arXiv:1602.08780 [stat.ML].

Tasche, D. (2017). Fisher consistency for prior probability shift. *Journal of Machine Learning Research*, 18:95:1–95:32.

Tasche, D. (2019). Confidence intervals for class prevalences under prior probability shift. *Machine Learning and Knowledge Extraction*, 1(3):805–831.

Tasche, D. (2021). Minimising quantifier variance under prior probability shift. arXiv:2107.08209 [stat.ML].

van Rijsbergen, C. J. (1979). *Information retrieval*. Butterworths, London, UK, second edition.

Vapnik, V. (1998). *Statistical learning theory*. Wiley, New York, US.

Viana, M. A., Ramakrishnan, V., and Levy, P. S. (1993). Bayesian analysis of prevalence from the results of small screening samples. *Communications in Statistics - Theory and Methods*, 22(2):575–585.

Vilalta, R., Giraud-Carrier, C., Brazdil, P., and Soares, C. (2011). Inductive transfer. In Sammut, C. and Webb, G. I., editors, *Encyclopedia of Machine Learning*, pages 545–548. Springer, Heidelberg, DE.

Vinyals, O., Bengio, S., and Kudlur, M. (2016). Order matters: Sequence to sequence for sets. In *Proceedings of the 4th International Conference on Learning Representations (ICLR 2016)*, San Juan, PR.

Vucetic, S. and Obradovic, Z. (2001). Classification on data with biased class distribution. In *Proceedings of the 12th European Conference on Machine Learning (ECML 2001)*, pages 527–538, Freiburg, DE.

Wagstaff, E., Fuchs, F., Engelcke, M., Posner, I., and Osborne, M. A. (2019). On the limitations of representing functions on sets. In *Proceedings of the 36th International Conference on Machine Learning (ICML 2019)*, pages 6487–6494, Long Beach, US.

Walker, M. A., Anand, P., Abbott, R., and Grant, R. (2012). Stance classification using dialogic properties of persuasion. In *Proceedings of the 2012 Conference of the North American Chapter of the Association for Computational Linguistics (HLT-NAACL 2012)*, pages 592–596, Montreal, CA.

Werman, M., Peleg, S., and Rosenfeld, A. (1985). A distance metric for multidimensional histograms. *Computer Vision, Graphics, and Image Processing*, 32:328–336.

Xiao, Y., Gordon, A., and Yakovlev, A. (2006). The L1-version of the Cramér-von Mises test for two-sample comparisons in microarray data analysis. *EURASIP Journal on Bioinformatics and Systems Biology*, 2006:1–9.

Xue, J. C. and Weiss, G. M. (2009). Quantification and semi-supervised classification methods for handling changes in class distribution. In *Proceedings of the 15th ACM International Conference on Knowledge Discovery and Data Mining (SIGKDD 2009)*, pages 897–906, Paris, FR.

Yang, C. and Zhou, J. (2008). Non-stationary data sequence classification using online class priors estimation. *Pattern Recognition*, 41(8):2656–2664.

Yang, Y. (2001). A study on thresholding strategies for text categorization. In *Proceedings of the 24th ACM International Conference on Research and Development in Information Retrieval (SIGIR 2001)*, pages 137–145, New Orleans, US.

Zadrozny, B. and Elkan, C. (2002). Transforming classifier scores into accurate multiclass probability estimates. In *Proceedings of the 8th ACM International Conference on Knowledge Discovery and Data Mining (KDD 2002)*, pages 694–699, Edmonton, CA.

Zaheer, M., Kottur, S., Ravanbakhsh, S., Poczos, B., Salakhutdinov, R. R., and Smola, A. J. (2017). Deep sets. In *Proceedings of the 31st Annual Conference on Neural Information Processing Systems (NIPS 2017)*, pages 3391–3401, Long Beach, US.

Zeiberg, D., Jain, S., and Radivojac, P. (2020). Fast nonparametric estimation of class proportions in the positive-unlabeled classification setting. In *Proceedings of the 34th AAAI Conference on Artificial Intelligence (AAAI 2020)*, pages 6729–6736, New York, US.

Zhang, Z. and Zhou, J. (2010). Transfer estimation of evolving class priors in data stream classification. *Pattern Recognition*, 43(9):3151–3161.

Zhou, X.-H., McClish, D. K., and Obuchowski, N. A. (2002). *Statistical methods in diagnostic medicine*. Wiley, New York, US.

Index

Printed in the United States
by Baker & Taylor Publisher Services